WORKBOOK FOR
MECHANICAL D

CAD-COMMUNICATIONS

Jay D. Helsel
Byron Urbanick

GLENCOE
Macmillan/McGraw-Hill

LAKE FOREST, ILLINOIS • COLUMBUS, OHIO • MISSION HILLS, CALIFORNIA • PEORIA, ILLINOIS

Workbook for Mechanical Drawing: CAD-Communications

Send all inquiries to:
Glencoe Division, Macmillan/McGraw-Hill
936 Eastwind Drive
Westerville, Ohio 43081

ISBN 0-07-022338-6

1 2 3 4 5 6 7 8 9 0 MALMAL 9 8 7 6 5 4 3 2 1 0

CONTENTS

PREFACE

This *Workbook for Mechanical Drawing: CAD-Communications* is correlated with the eleventh edition of the drafting text, *Mechanical Drawing: CAD-Communications* by French, Svensen, Helsel, and Urbanick. Although many of the drawing assignments are taken directly from the text and references in several of the assignments are directed to the text, this workbook can easily be used with any good text that follows current ANSI standards. The instructor simply will need to provide the text reference when necessary.

We recognize the fast-growing use of computer graphics and have made every effort to provide an appropriate balance of traditional drawing (TRAD) assignments and computer-aided drawing (CAD) assignments. In fact, nearly all the assignments can be used equally well with either TRAD or CAD. However, we believe that the students need to develop a good working knowledge of the fundamentals of projection, dimensioning, and other basic elements of technical graphics as a communications subject before working entirely on a computerized system. This workbook provides complete flexibility for teachers to develop the mix of TRAD and CAD that is most appropriate for their individual programs.

Sufficient drawing assignments are included to thoroughly cover all basic elements of drafting. In addition, a variety of advanced problems are provided to accommodate individual differences in students and programs.

Since the organization of the workbook follows that of the textbook it is relatively easy for students to progress at their own best rate. For those who progress more quickly than others, a quantity of sheets with only a border and title strip are included for advanced work. These sheets may also be used for remedial work throughout the course. In some of the highly specialized units, a rather limited number of assignments are provided. For those who wish to expand those units, problems may be assigned directly from the textbook and drawn on the supplied bordered sheets to provide uniformity throughout the student's set of drawings.

Considerable time has been spent in developing and testing the material used in this book. Care was taken in the selection of drawing problems to ensure a broad range of experiences that follow ANSI standards and to provide students with a solid background in the elements of drafting as a communications subject.

We are grateful to the many instructors who have used *Mechanical Drawing* over its very successful history. We hope that those instructors and others will find the workbook as useful as the textbook. As always, we appreciate your comments and suggestions.

Jay D. Helsel
Byron Urbanick

Study the illustrations in the Preface and Chapter 1 of your textbook. For each of the figure numbers listed below, give the branch of engineering represented and the type of activity involved in each. Letter your answers neatly in single-stroke vertical capital letters as though you were making notes for your design director.

EXAMPLE:

FIGURE NO.	BRANCH OF ENGINEERING	MAJOR ACTIVITY
1A	ELECTRONIC DESIGN	COMPUTER MICROCHIP LAYOUT

FIGURE NO.	BRANCH OF ENGINEERING	MAJOR ACTIVITY
1-2		
1-3		
1-4		
1-6B		
1-7		
1-10A		
1-10B		
1-11		
1-12		
1-20A		
1-20B		
1-21		
1-24		
1-25		
1-26		
1-27		
1-30		
1-31		
1-33		
1-48		

List the five major branches of engineering and give several major functions and activities of each. Letter your answers carefully and neatly in the space provided below.

1.

2.

3.

4.

5.

TWENTY-FIRST CENTURY DRAFTING	DRAWN BY	DATE	DWG NO.

DRAWN BY
DATE
DWG NO.

COMPUTER-AIDED DRAFTING . . . C A D – T R A D . . . TRADITIONAL DRAFTING

Study Chapter 1 in your textbook and complete the following assignments. Letter your answers neatly in single-stroke vertical capital letters.

A. List twelve major characteristics of a CAD system. Limit each response to one line of lettering.

1.
2.
3.
4.
5.
6.
7.
8.
9.
10.
11.
12.

B. Name six major components of a CAD turnkey system.

1.
2.
3.
4.
5.
6.

C. List six major components of a traditional drafting system.

1.
2.
3.
4.
5.
6.

TWENTY-FIRST CENTURY DRAFTING	DRAWN BY	DATE	DWG NO.

DRAWN BY
DATE
DWG NO.

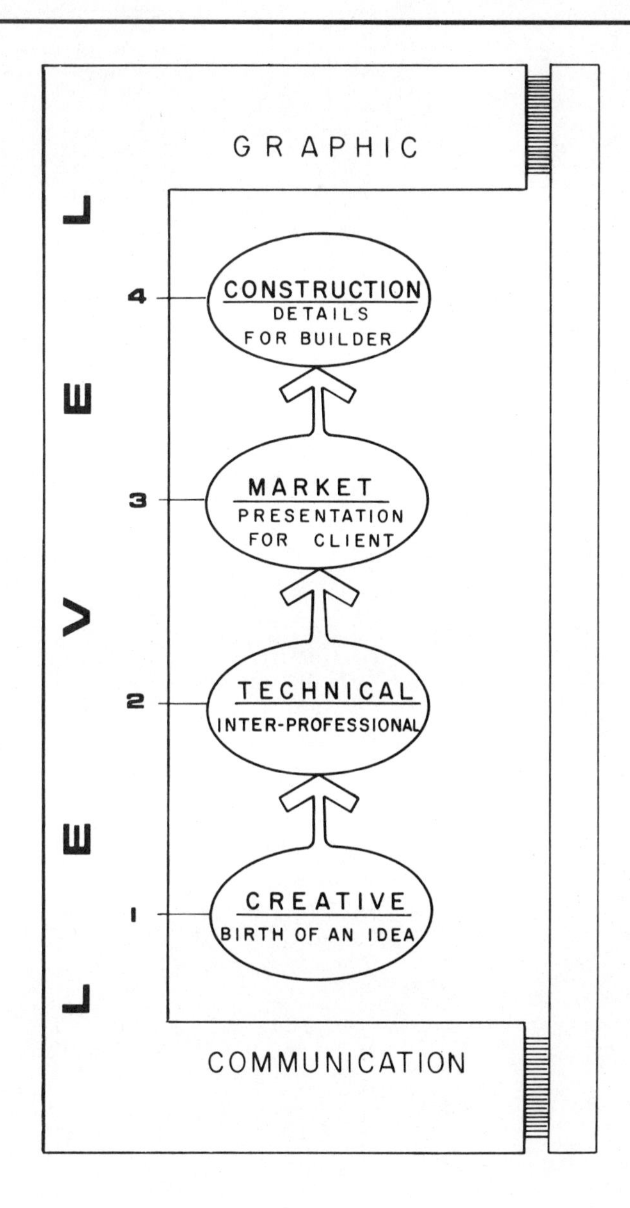

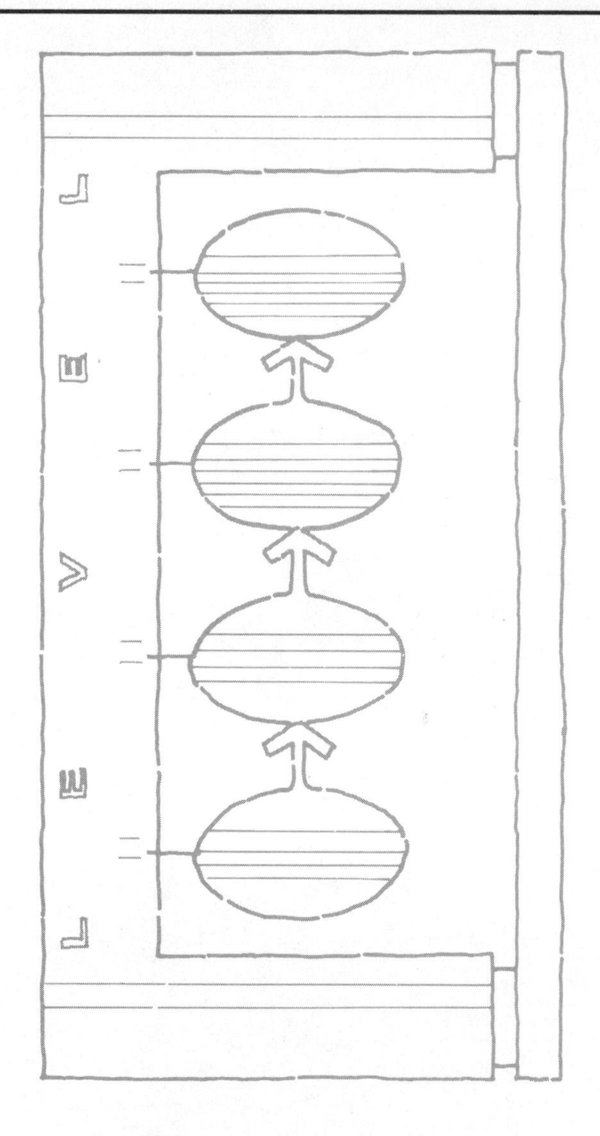

Use good line technique to darken the lines in the chart above. Complete the chart by lettering all information in the guidelines given. Do not darken the guidelines.

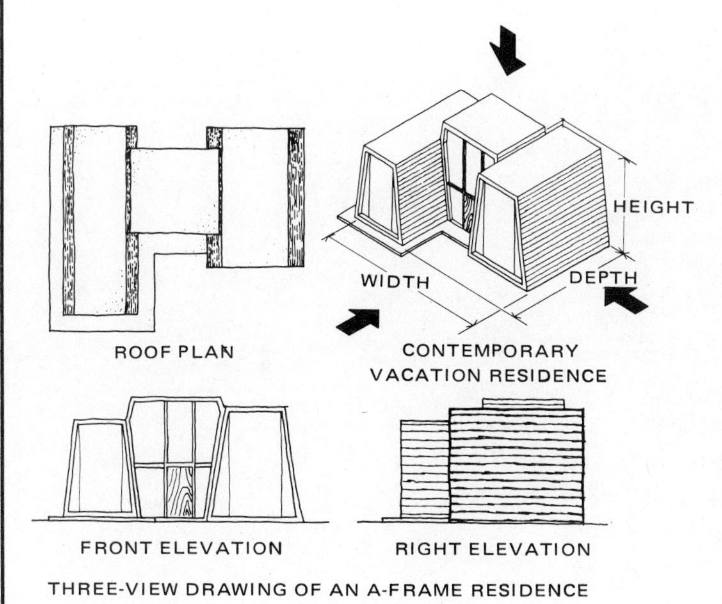

THREE-VIEW DRAWING OF AN A-FRAME RESIDENCE

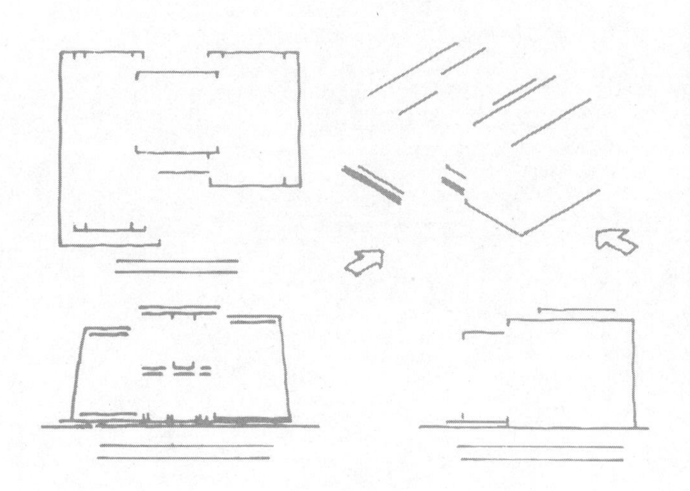

Complete the sketch of the *A-frame residence.*
Use good sketching technique as you darken the gray lines.

TWENTY-FIRST CENTURY DRAFTING	DRAWN BY	DATE	DWG NO.

DRAWN BY
DATE
DWG NO.

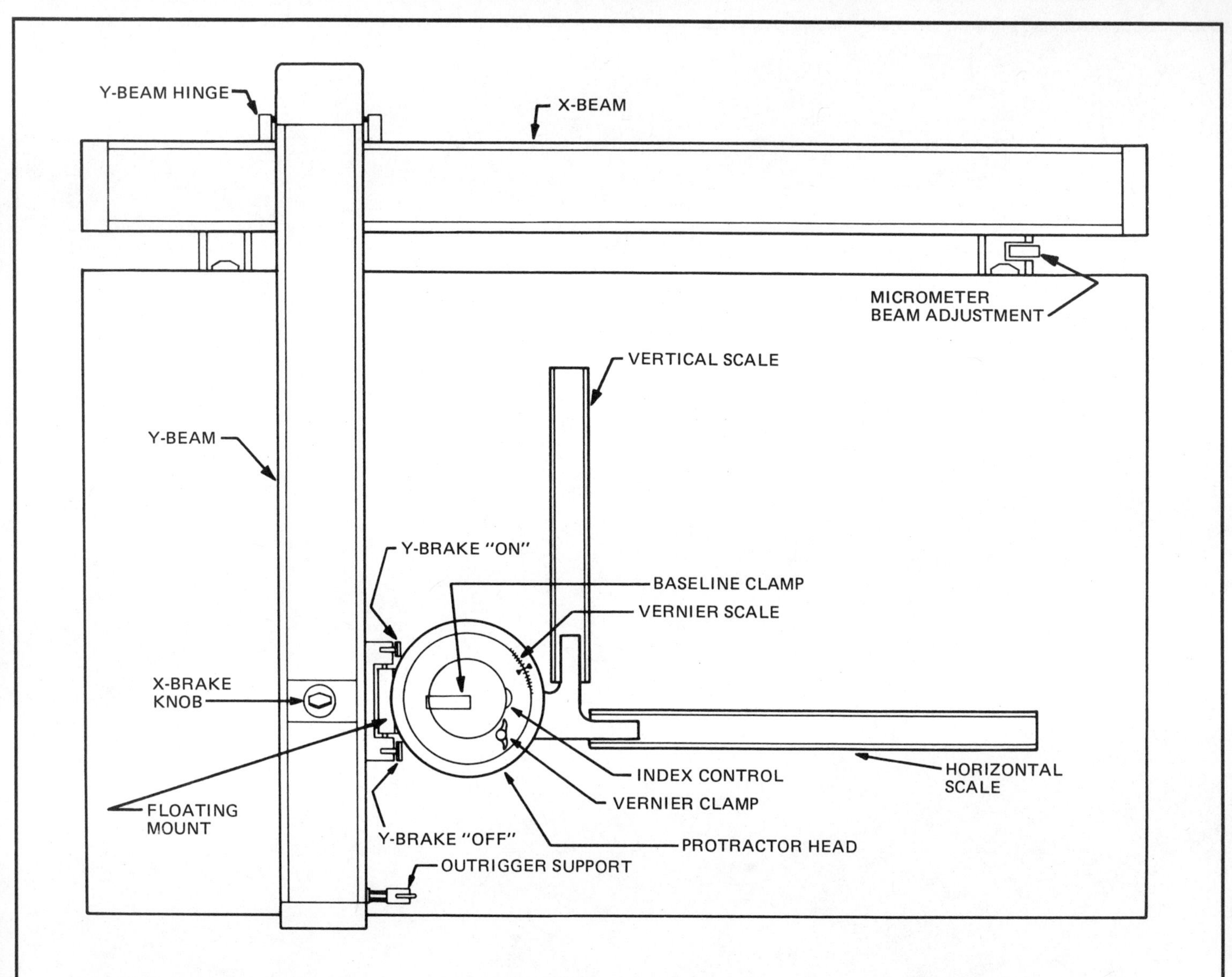

Listed below are many of the parts of the track drafter shown above. To the right of each item, letter the CAD option similar in function to each corresponding part of the track drafter. Letter your answers neatly in single-stroke vertical capital letters.

X-BEAM ________________

Y-BEAM ________________

PROTRACTOR HEAD ________________

VERTICAL SCALE ________________

HORIZONTAL SCALE ________________

BASELINE CLAMP ________________

VERNIER CLAMP ________________

INDEX CONTROL ________________

FLOATING MOUNT ________________

X-BRAKE ________________

Y-BRAKE ________________

TWENTY-FIRST CENTURY DRAFTING	DRAWN BY	DATE	DWG NO.

	DRAWN BY	DATE	DWG NO.

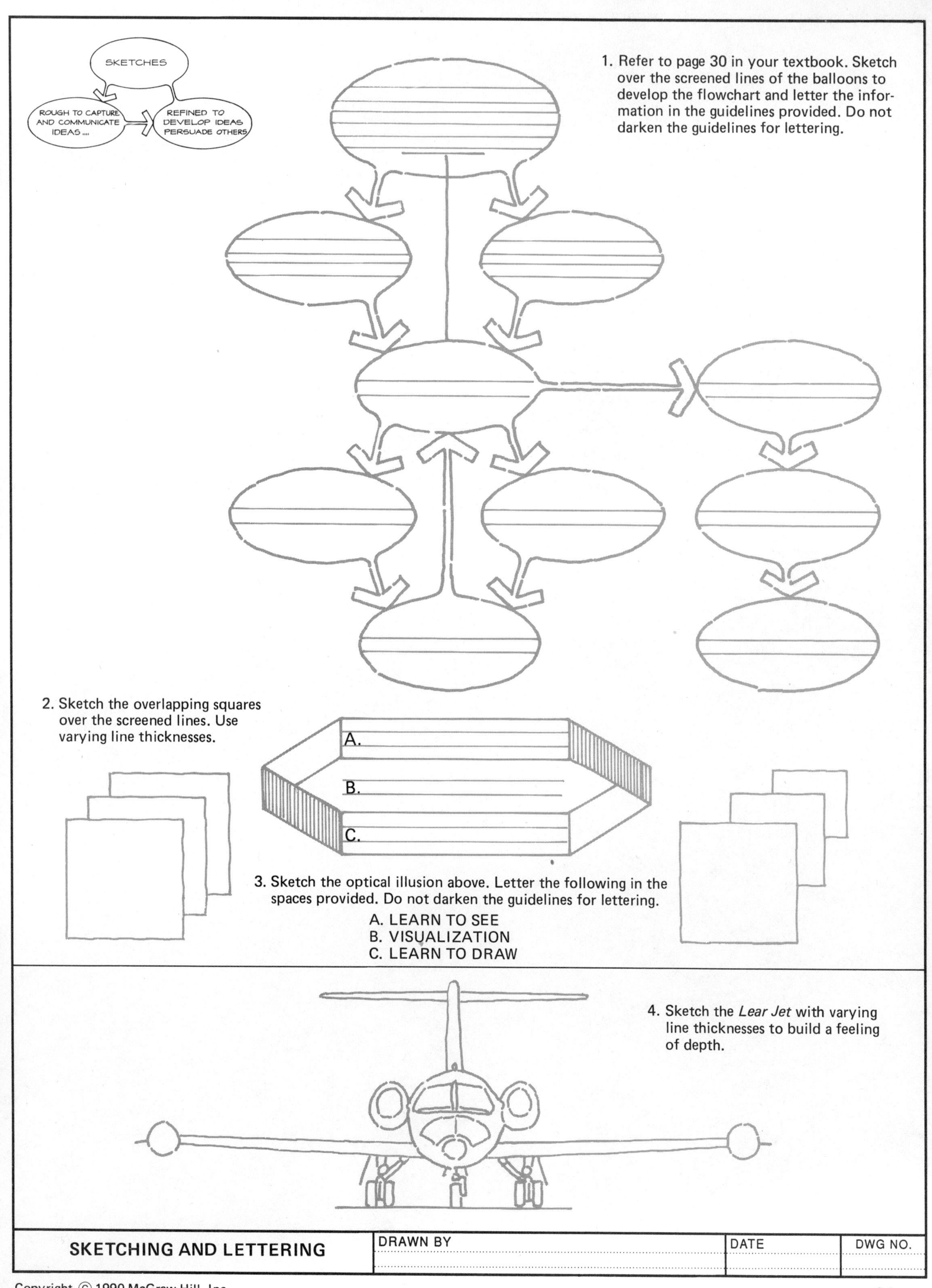

1. Refer to page 30 in your textbook. Sketch over the screened lines of the balloons to develop the flowchart and letter the information in the guidelines provided. Do not darken the guidelines for lettering.

2. Sketch the overlapping squares over the screened lines. Use varying line thicknesses.

3. Sketch the optical illusion above. Letter the following in the spaces provided. Do not darken the guidelines for lettering.
   A. LEARN TO SEE
   B. VISUALIZATION
   C. LEARN TO DRAW

4. Sketch the *Lear Jet* with varying line thicknesses to build a feeling of depth.

SKETCHING AND LETTERING	DRAWN BY	DATE	DWG NO.

DRAWN BY
DATE
DWG NO.

Sketch the six-balloon diagram over the screened lines. Letter the information given in the diagram below in the guidelines provided. Do not darken the guidelines.

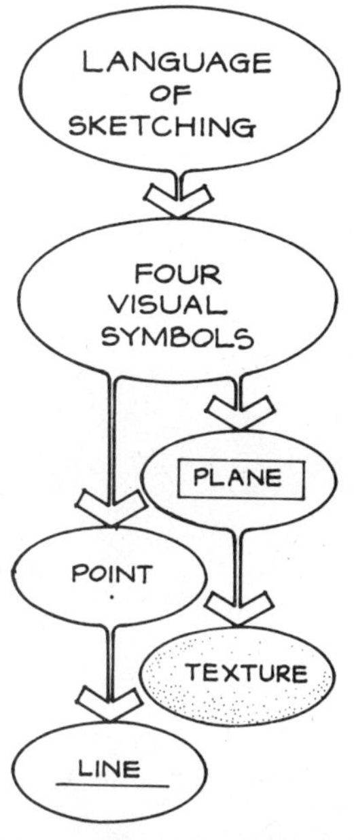

The elements of sketching are simple visual symbols.

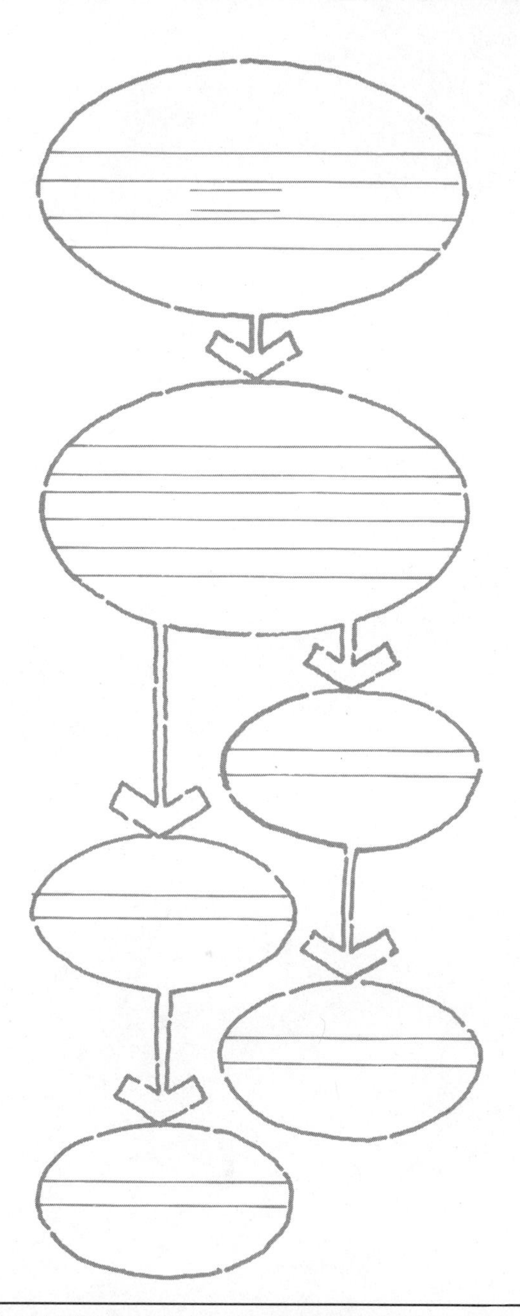

Use lines to show how you can control wireframe cubes A, B, and C.

A

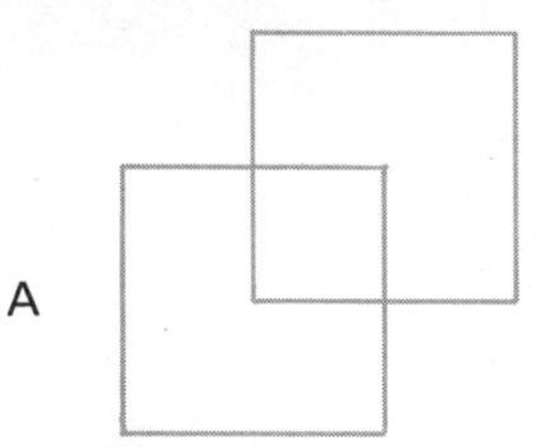

Which square tends to stay in front?

B

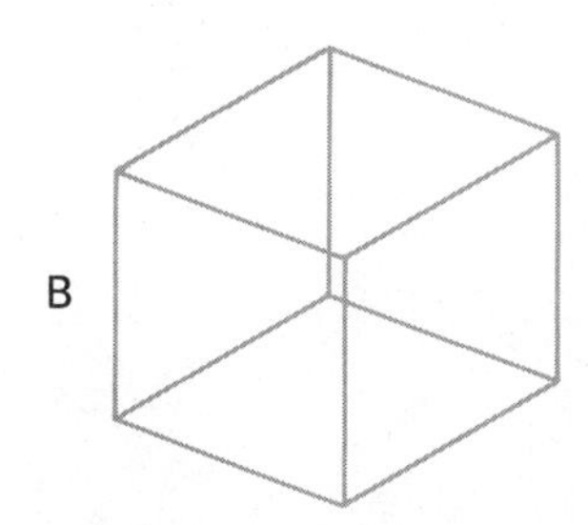

Can the top view dominate? Do the top and bottom views fluctuate readily?

C

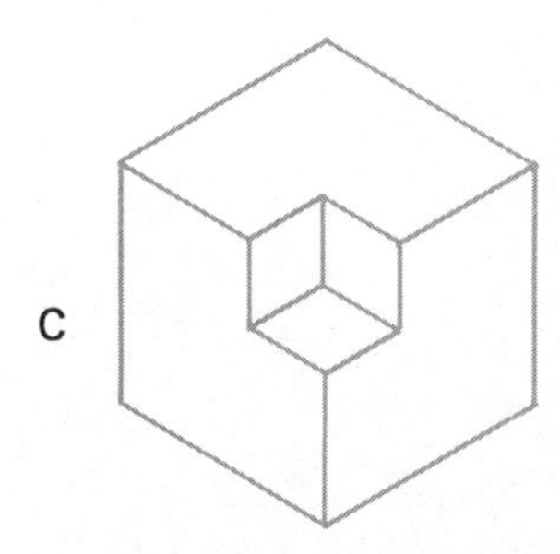

Is the small cube projecting from, or cut out of, the large cube?

OCTAHEDRON IN A TRANSPARENT CUBE

BUILD VISUAL POWER

Construct an octahedron in the transparent cube at the right.

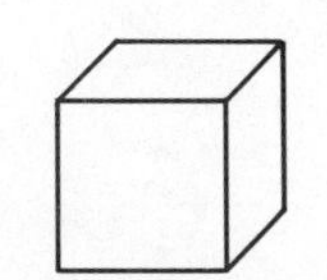

1. Create a 1.25″ oblique (cabinet) cube.

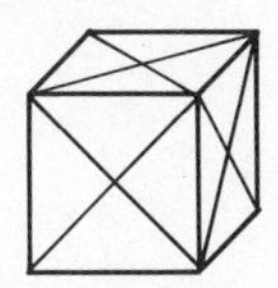

2. Locate the center of the six sides.
3. Connect these points with diagonals.

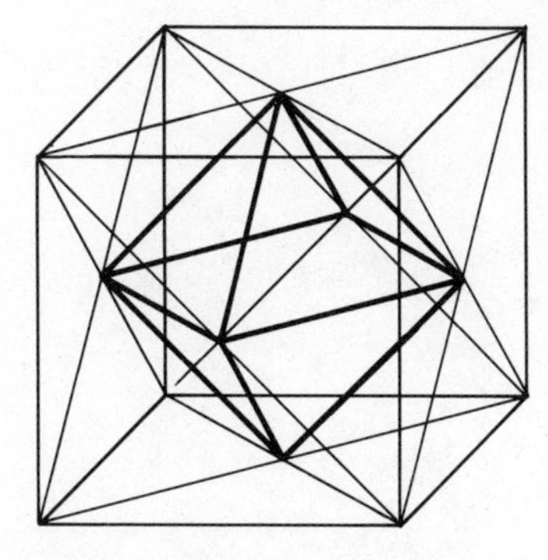

Transparent cube with diagonals and octahedron.

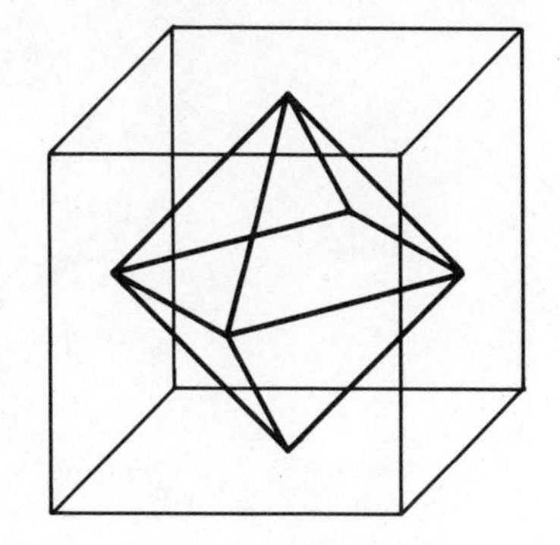

Diagonals removed.

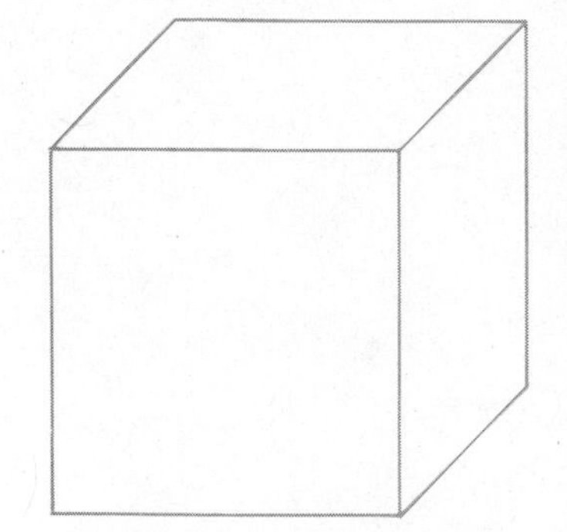

Sketch the transparent cube with the octahedron. Use contrasting line weights to emphasize the octahedron.

SKETCHING AND LETTERING	DRAWN BY	DATE	DWG NO.

DRAWN BY
DATE
DWG NO.

Study the transparent forms. In the blank shapes below, create eight different designs. Use shaded areas to develop three-dimensional images as shown in the examples. Develop your designs using freehand sketching only.

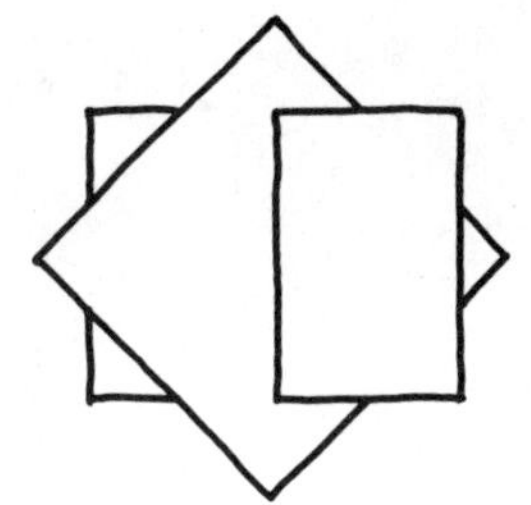
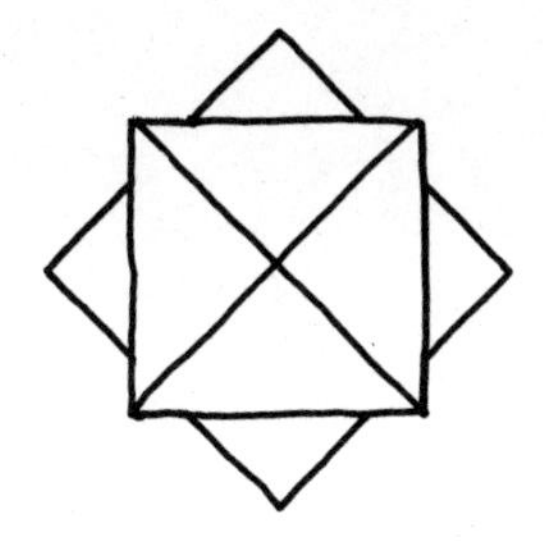
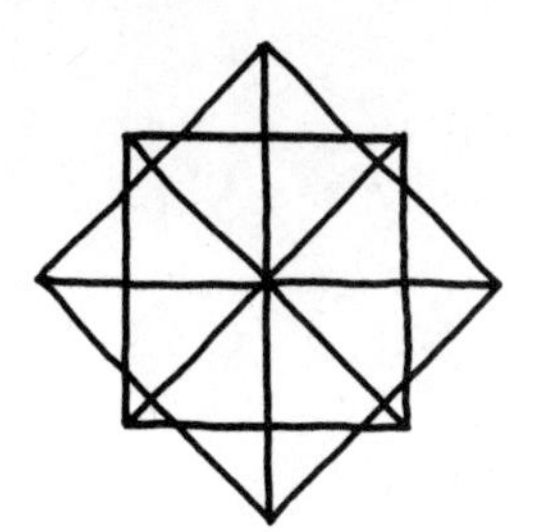

TRANSPARENT FORMS

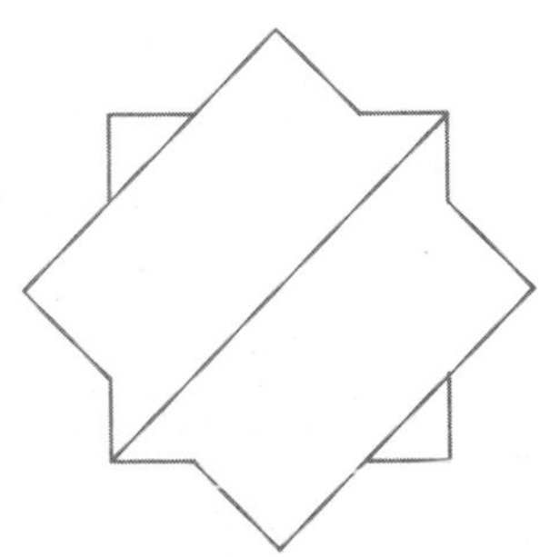
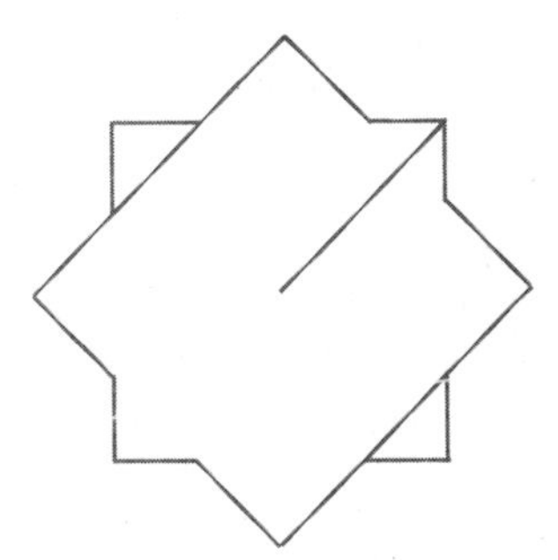

EXAMPLES

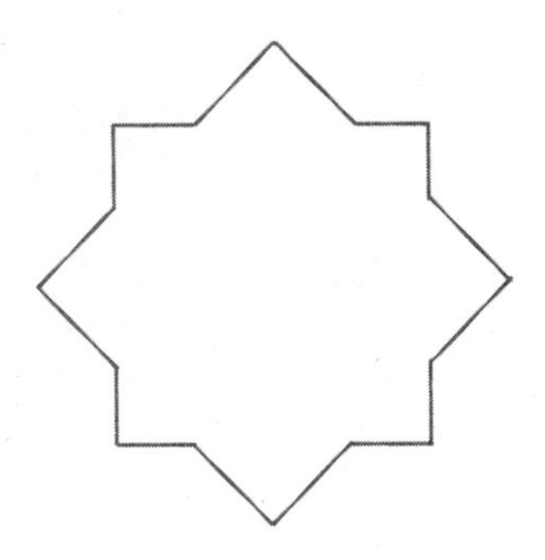
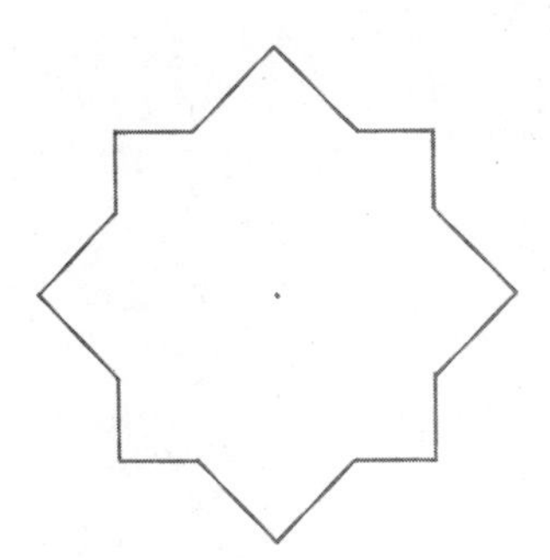

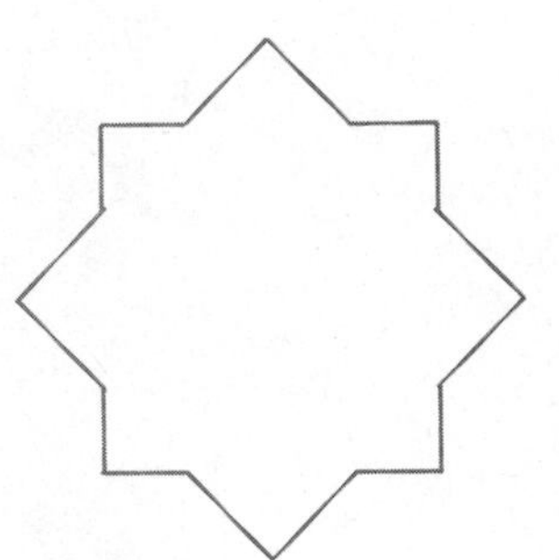
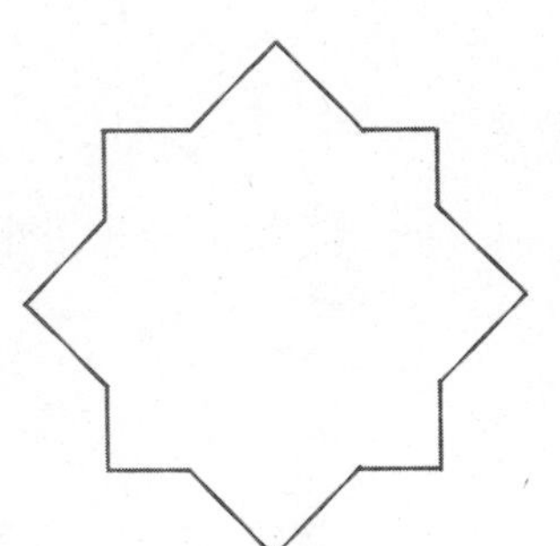
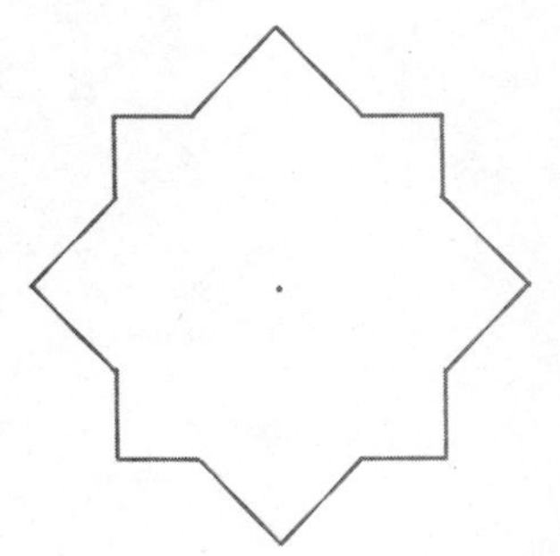

BUILDING VISUAL POWERS WITH MONDRIAN FORMS

SKETCHING	DRAWN BY	DATE	DWG NO.

	DRAWN BY	DATE	DWG NO.

Sketch over each of the screened leader lines and add arrowheads. Carefully and skillfully letter the names of each of the lines using the corresponding guidelines. Do not darken guidelines. This assignment will help you learn to identify and use the types of lines recommended by ANSI (American National Standards Institute).

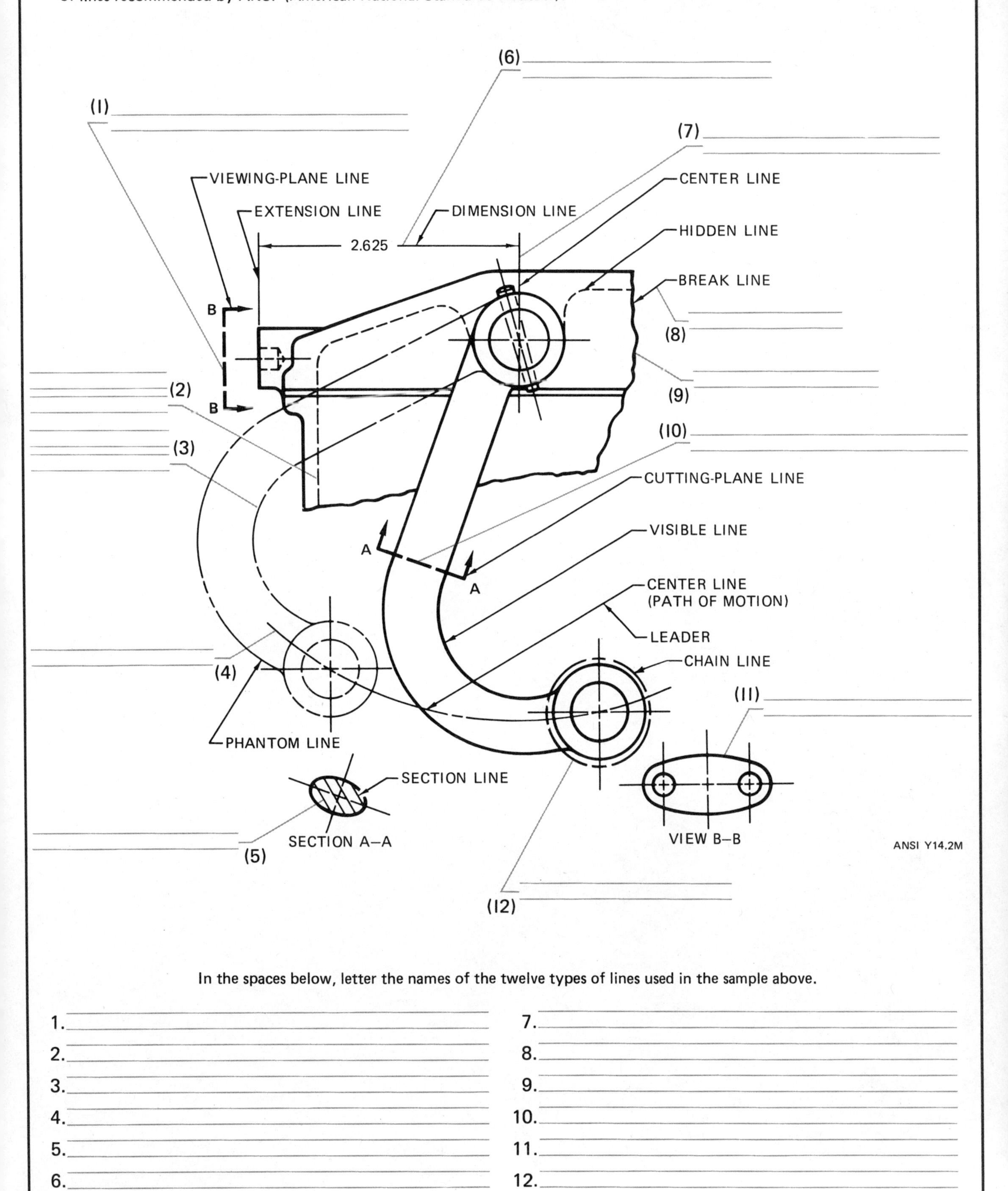

In the spaces below, letter the names of the twelve types of lines used in the sample above.

1.
2.
3.
4.
5.
6.
7.
8.
9.
10.
11.
12.

SKETCHING AND LETTERING	DRAWN BY	DATE	DWG NO.

DRAWN BY
DATE
DWG NO.

COLUMN 1

Using pencil, sketch additional lines that match the density and width of the printed samples below.

VISIBLE LINES

HIDDEN LINES

SECTION LINES

CENTERLINES

DIMENSION AND EXTENSION LINES

6.38

CUTTING-PLANE LINES

ALTERNATE CUTTING-PLANE LINES

BREAK LINES

BREAK LINES

PHANTOM LINES

COLUMN 2

Using the alphabet of lines from your CAD system, sketch or draw lines in Column 2 that correspond to the types of lines in Column 1.

SKETCHING AND LETTERING	DRAWN BY	DATE	DWG NO.

DRAWN BY
DATE
DWG NO.

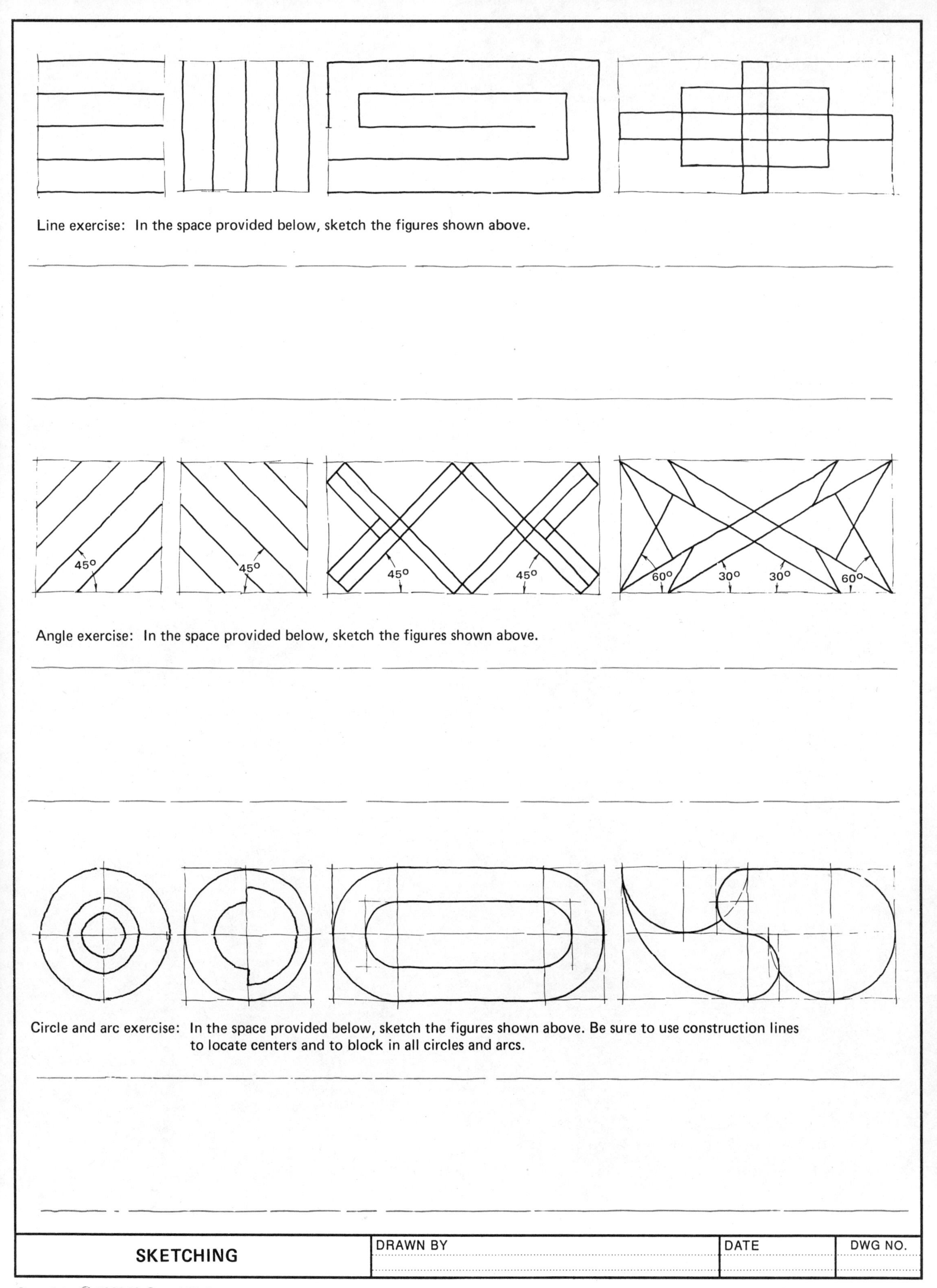

Line exercise: In the space provided below, sketch the figures shown above.

Angle exercise: In the space provided below, sketch the figures shown above.

Circle and arc exercise: In the space provided below, sketch the figures shown above. Be sure to use construction lines to locate centers and to block in all circles and arcs.

SKETCHING	DRAWN BY	DATE	DWG NO.

DRAWN BY
DATE
DWG NO.

Sketch the front, top, and right-side views of each of the pictorial objects shown. One corner of each view is marked. One block on the pictorial grid is equal to one block on the three-view drawing.

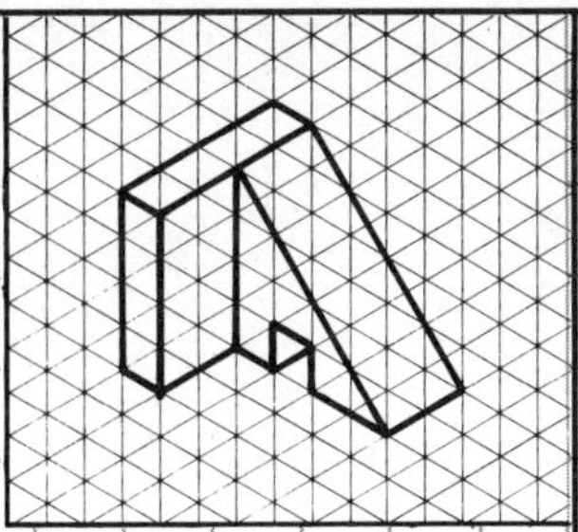

WEDGE BLOCK

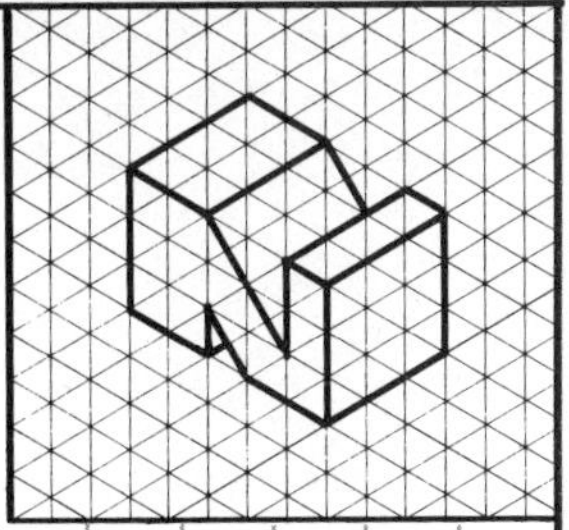

N-BLOCK

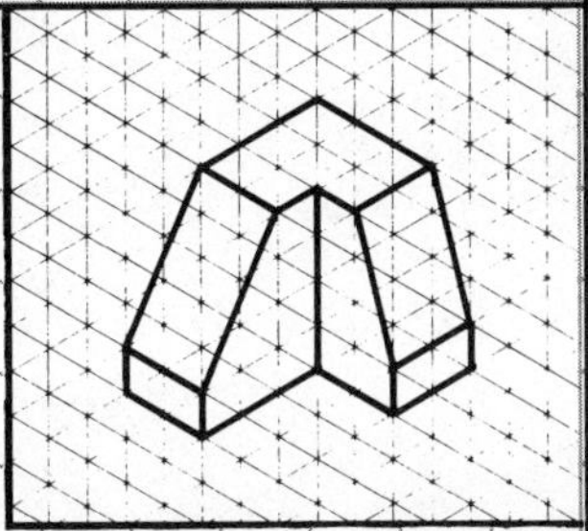

CORNER BLOCK

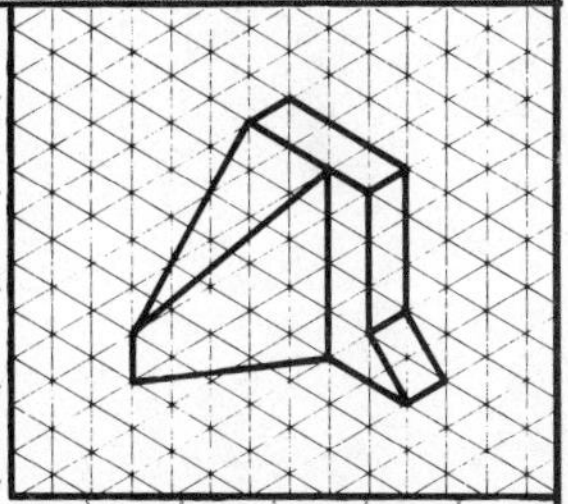

SPACER BLOCK

SKETCHING	DRAWN BY	DATE	DWG NO.

DRAWN BY
DATE
DWG NO.

Sketch a right-side view of the four objects shown. Sketch an isometric pictorial on the grid provided.

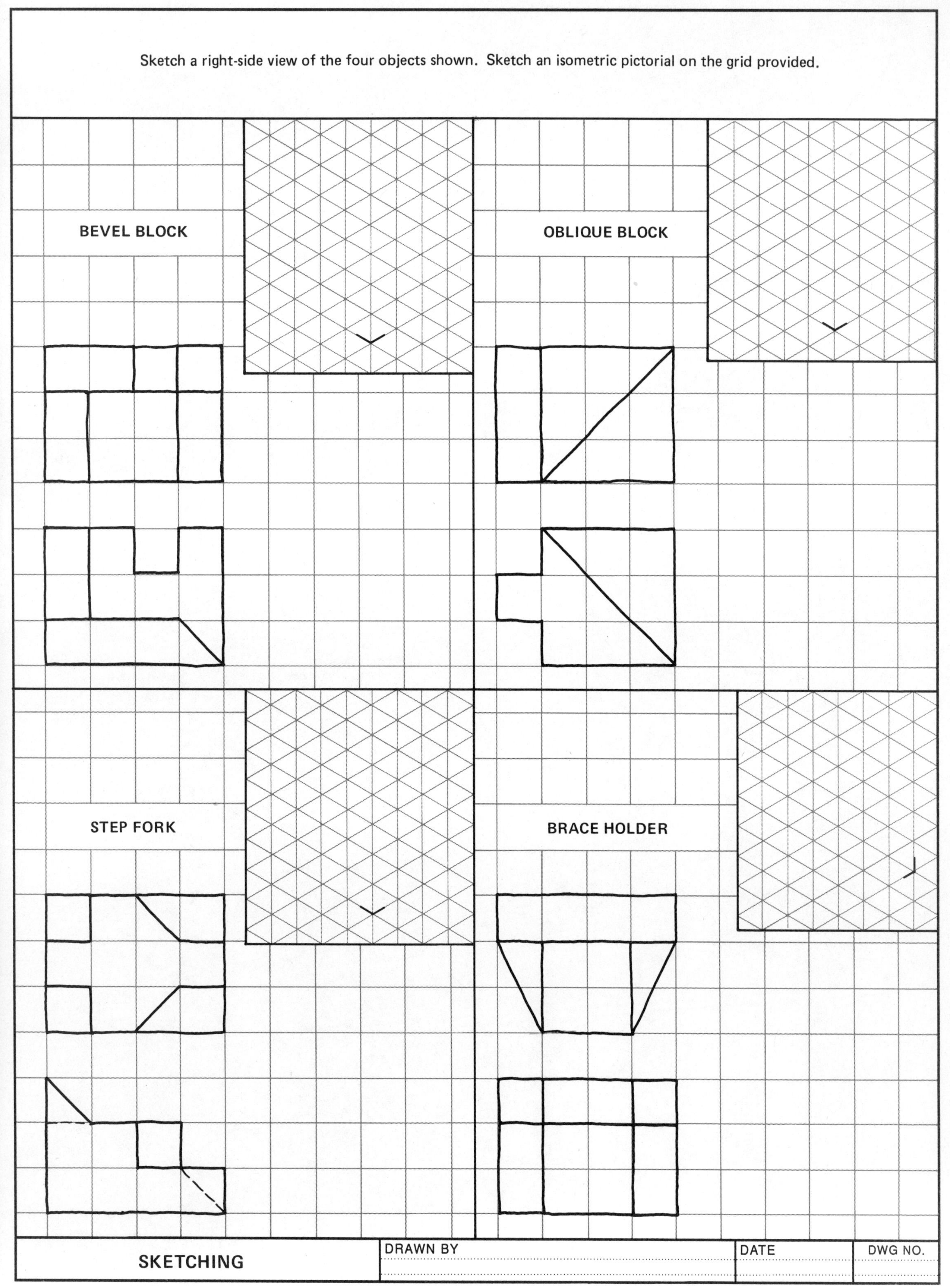

SKETCHING	DRAWN BY	DATE	DWG NO.

	DRAWN BY	DATE	DWG NO.

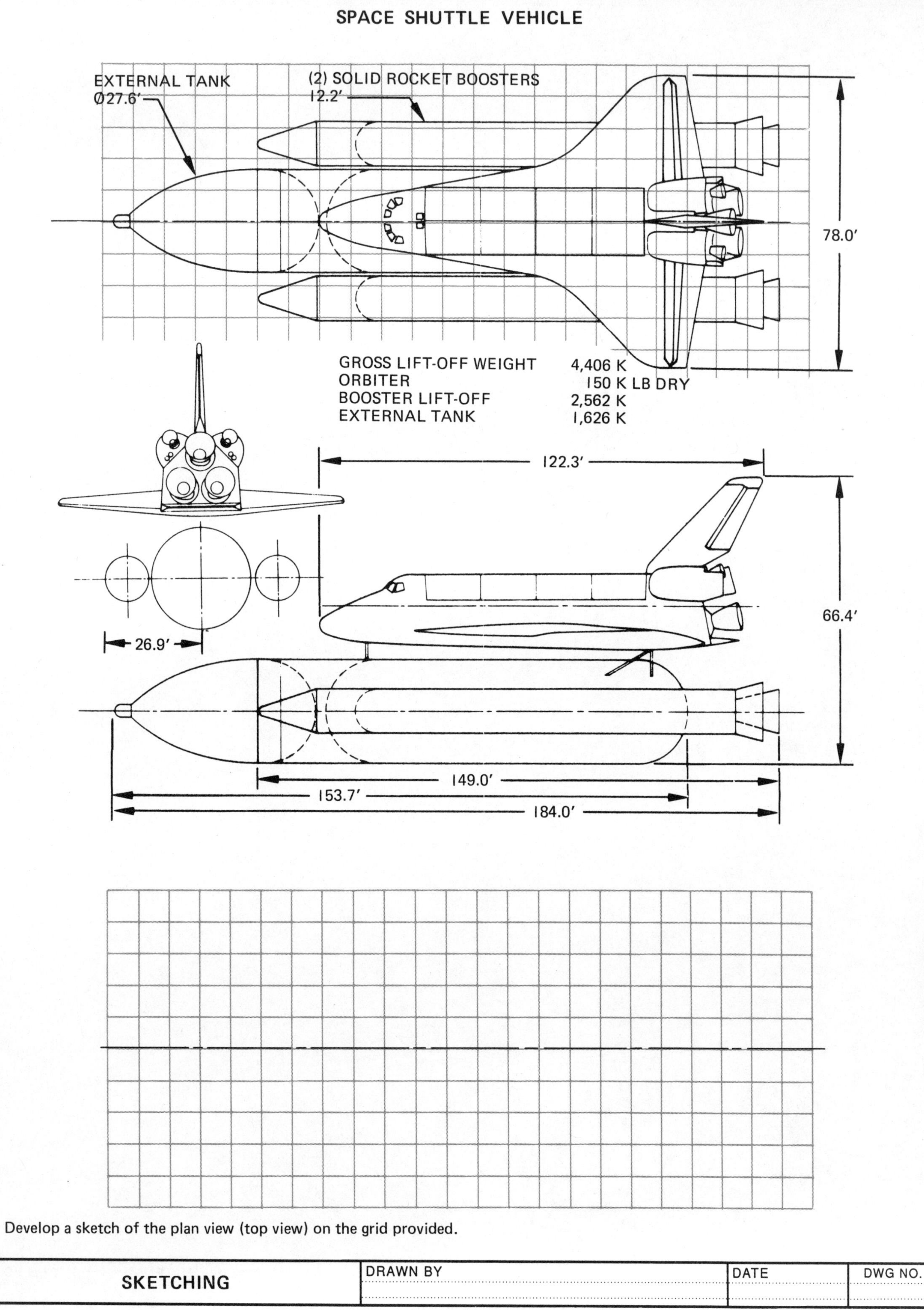

Develop a sketch of the plan view (top view) on the grid provided.

SKETCHING	DRAWN BY	DATE	DWG NO.

DRAWN BY
DATE
DWG NO.

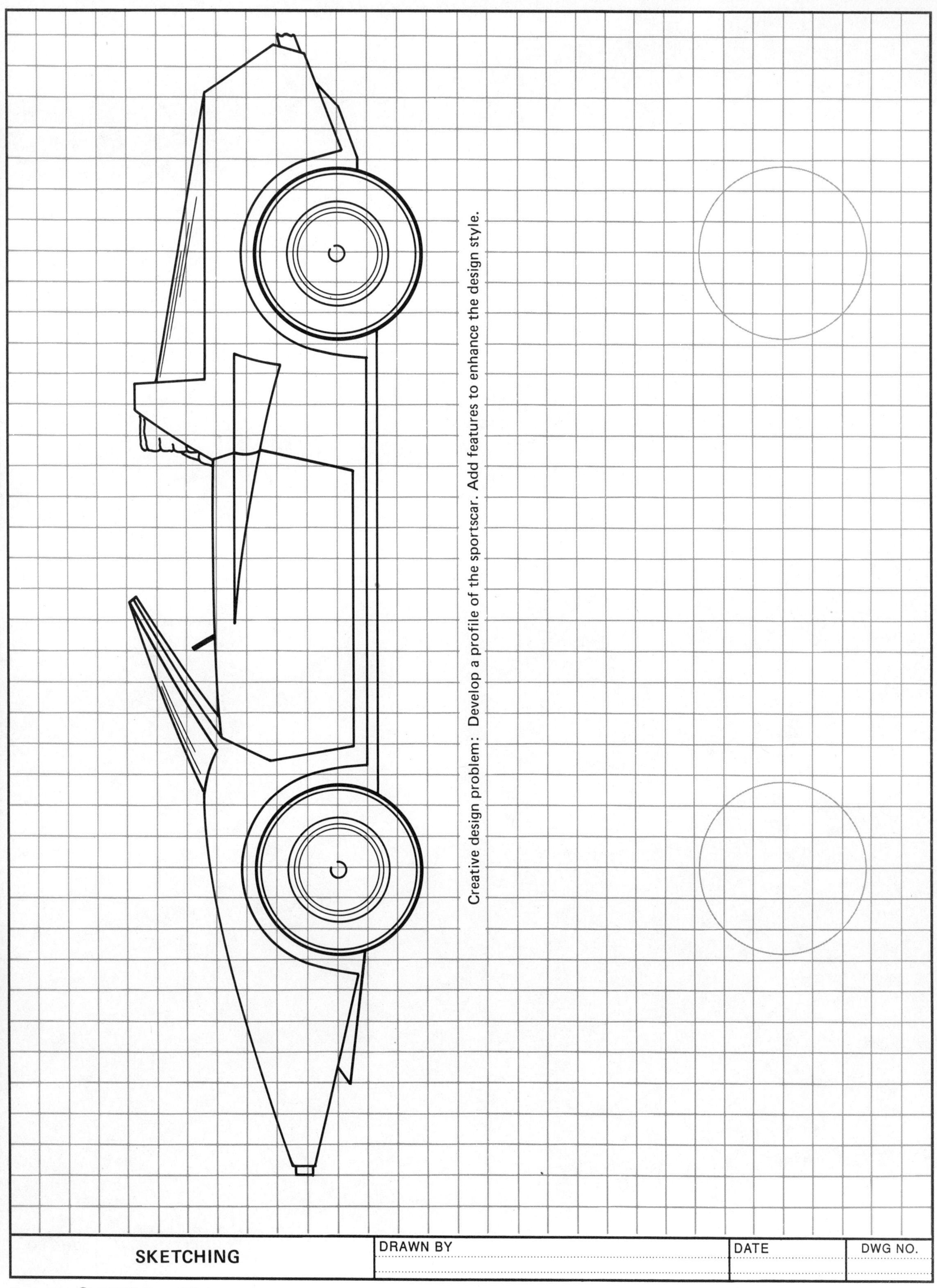
Creative design problem: Develop a profile of the sportscar. Add features to enhance the design style.
SKETCHING
DRAWN BY
DATE
DWG NO.

	DRAWN BY	DATE	DWG NO.

THE ALPHABET OF LINES (ANSI)

The illustrations below are for your information while completing the assignments on lines and line weights (thicknesses).

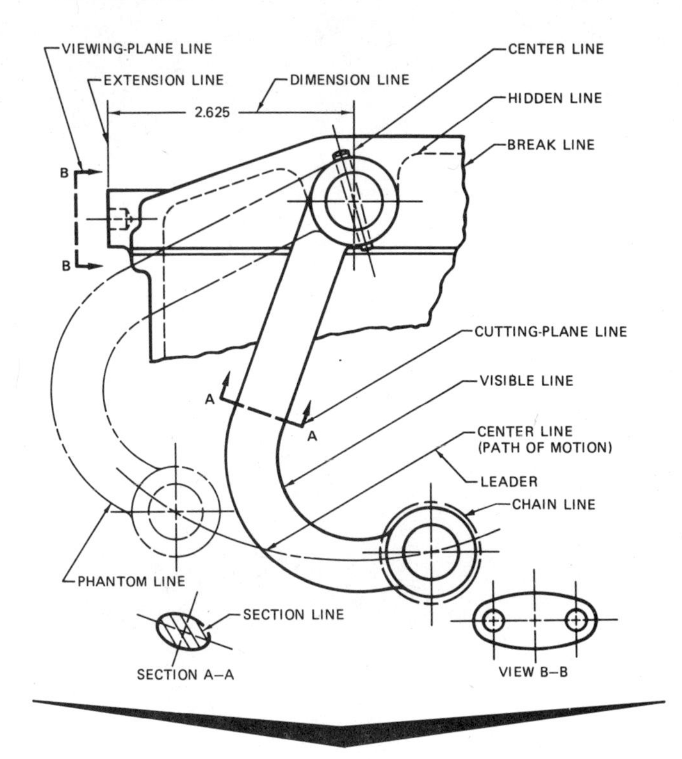

THICK — VISIBLE LINE

THIN — HIDDEN LINE

THIN — SECTION LINE

THIN — CENTER LINE OR SYMMETRY LINE

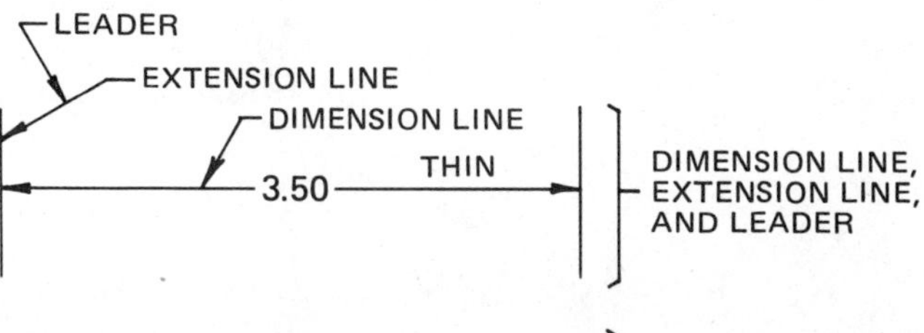

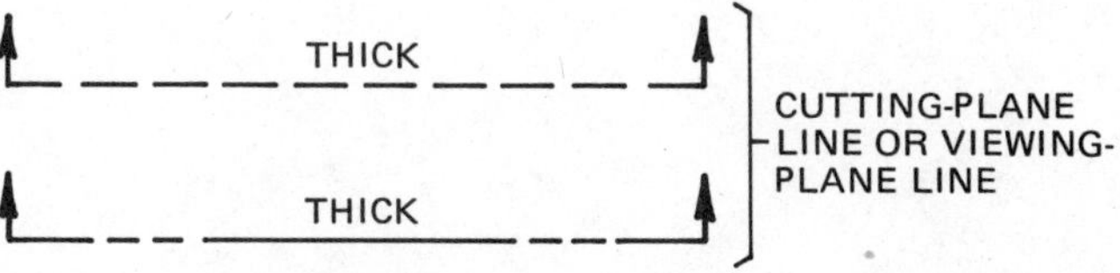

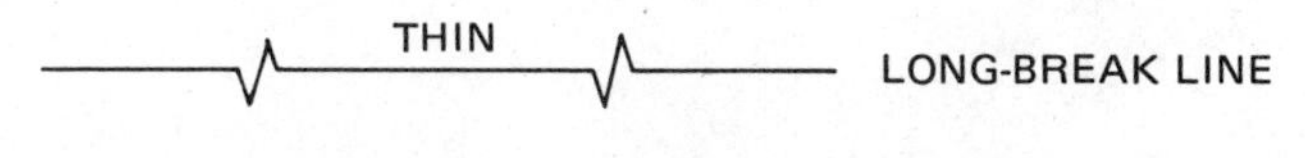

ASSIGNMENT

Using pencil, draw additional lines that match the density and width of the printed samples shown below.

VISIBLE LINES

HIDDEN LINES

SECTION LINES

CENTER LINES

DIMENSION LINES

$3\frac{1}{2}$

CUTTING-PLANE LINES

VIEWING-PLANE LINES

CUTTING-PLANE LINES

VIEWING-PLANE LINES

SHORT-BREAK LINES

LONG-BREAK LINES

PHANTOM LINES

THE USE AND CARE OF DRAFTING EQUIPMENT	DRAWN BY	DATE	DWG NO.

	DRAWN BY	DATE	DWG NO.

LEARNING TO DRAW LINES

Draw additional lines in each of the spaces below. In each case, match the given line in width and appearance. Draw all lines in the direction shown by the arrow (opposite direction for left-handed students).

HORIZONTAL	VERTICAL
INCLINED 45°	INCLINED 45°
INCLINED 30°	INCLINED 30°

THE USE AND CARE OF DRAFTING EQUIPMENT	DRAWN BY	DATE	DWG NO.

	DRAWN BY	DATE	DWG NO.

LEARNING TO DRAW LINES

Draw additional lines in each of the spaces below. In each case, match the given line in width and appearance. Draw all lines in the direction shown by the arrow (opposite direction for left-handed students).

INCLINED 60°	INCLINED 15°
INCLINED 60°	INCLINED 75°
INCLINED 15°	INCLINED 75°

THE USE AND CARE OF DRAFTING EQUIPMENT	DRAWN BY	DATE	DWG NO.

DRAWN BY
DATE
DWG NO.

Determine the length of each of the lettered dimension lines on the scale below. Letter each dimension in the corresponding spaces at the right. Letter carefully within the guidelines. Letter A is given as an example. Add arrowheads to the ends of the dimension lines.

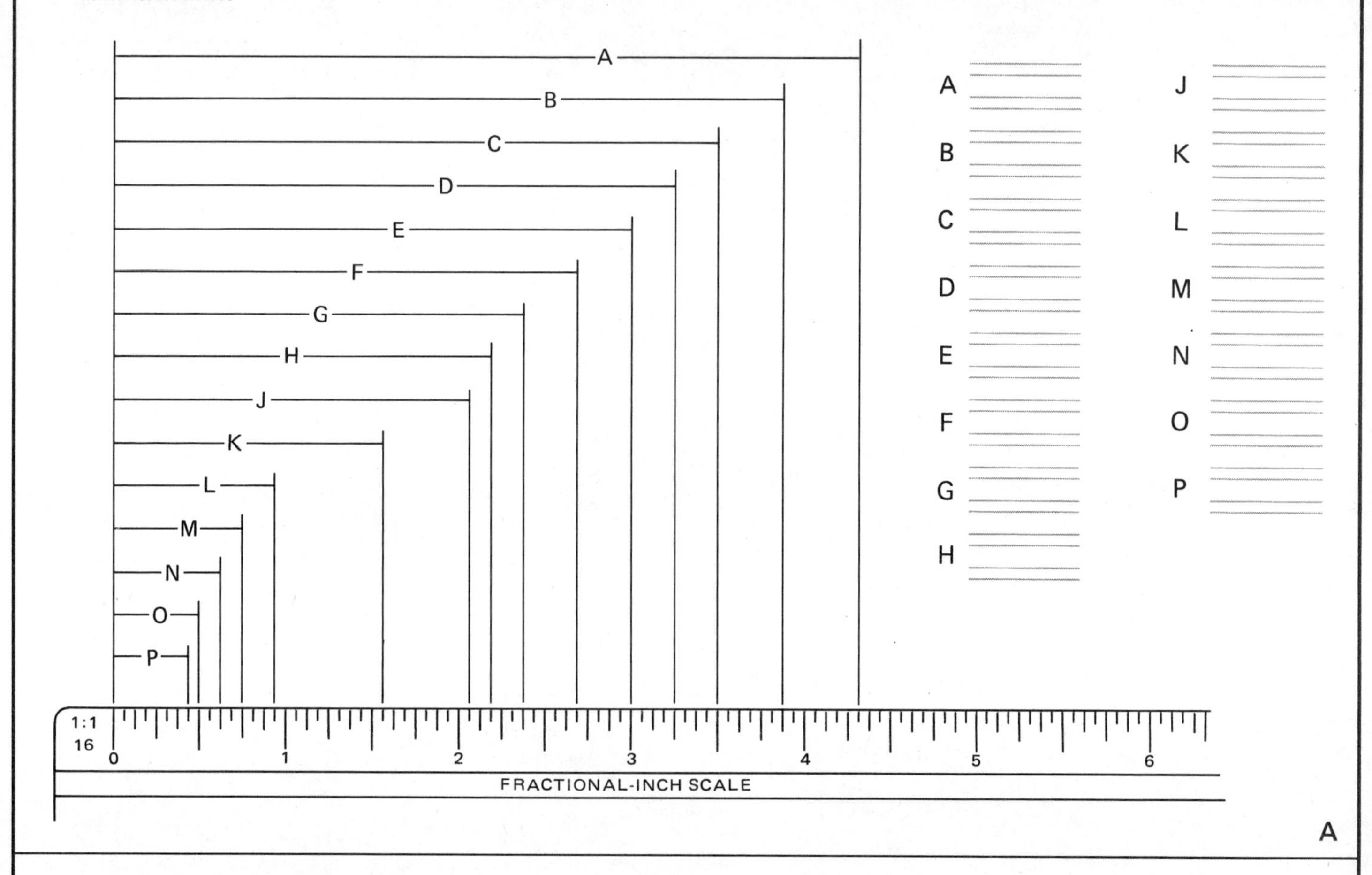

A

Using a full-size fractional-inch scale, measure each of the lettered lines in the figure below. Record the length of each in its corresponding space at the right. Measure all lines to the nearest 16th of an inch.

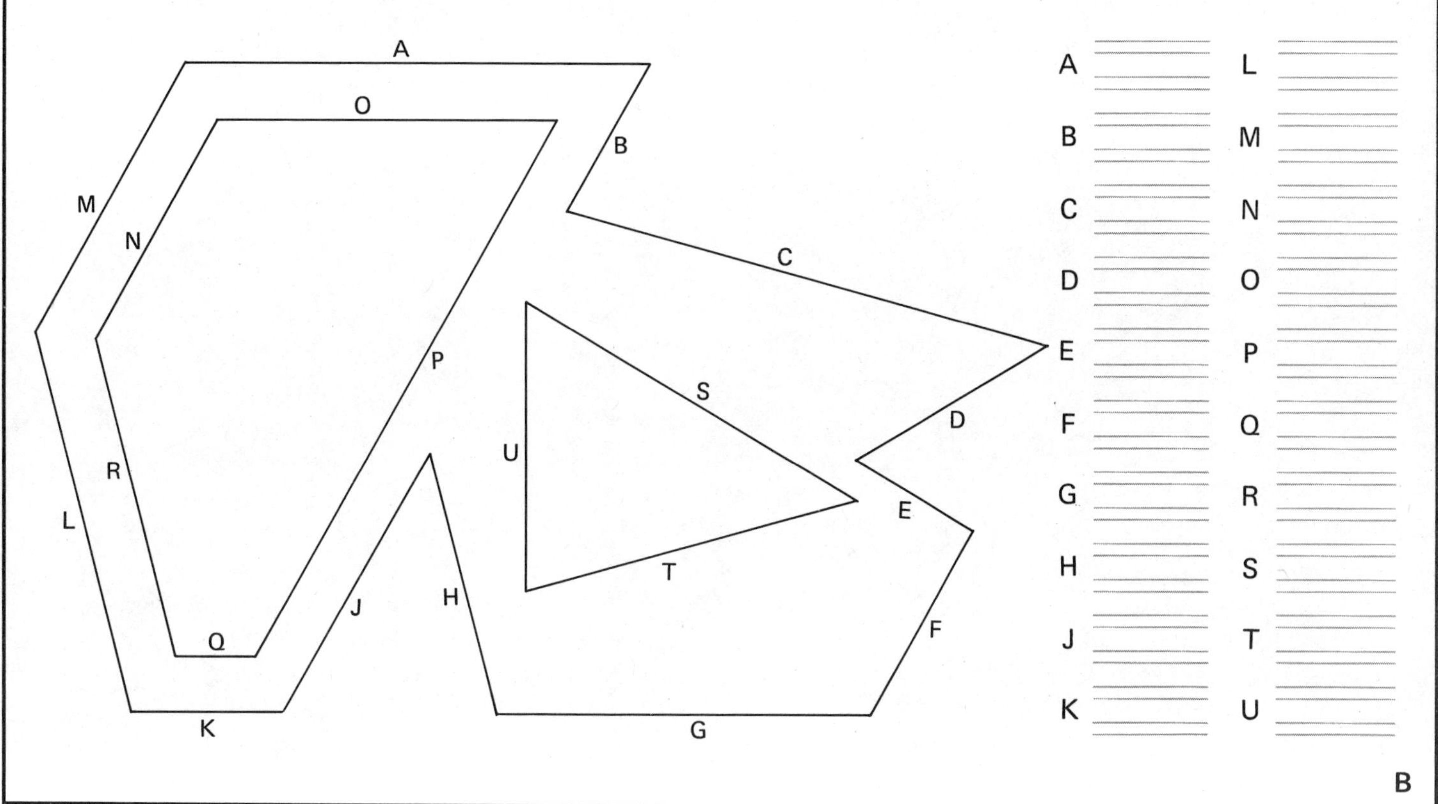

B

THE USE AND CARE OF DRAFTING EQUIPMENT	DRAWN BY	DATE	DWG NO.

	DRAWN BY	DATE	DWG NO.

Determine the length of each dimension on the architect's scale. Record each in the corresponding lettered spaces below. Give all answers in feet and inches. Add arrowheads to the ends of the dimension lines. Check your answers carefully.

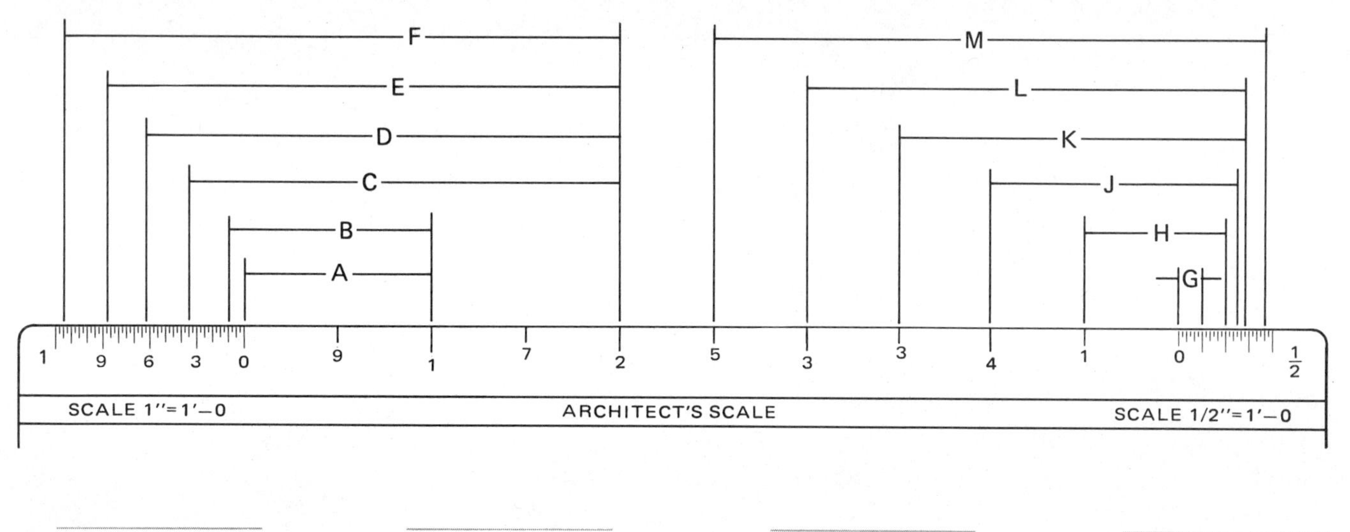

A ______ D ______ G ______ K ______

B ______ E ______ H ______ L ______

C ______ F ______ J ______ M ______

A

Use a 1/4"=1'–0 architect's scale to measure each of the dimensions below. Place the dimension above each dimension line as shown in the 20'–0 dimension. Measure accurately and give all dimensions to the nearest inch. Add arrowheads.

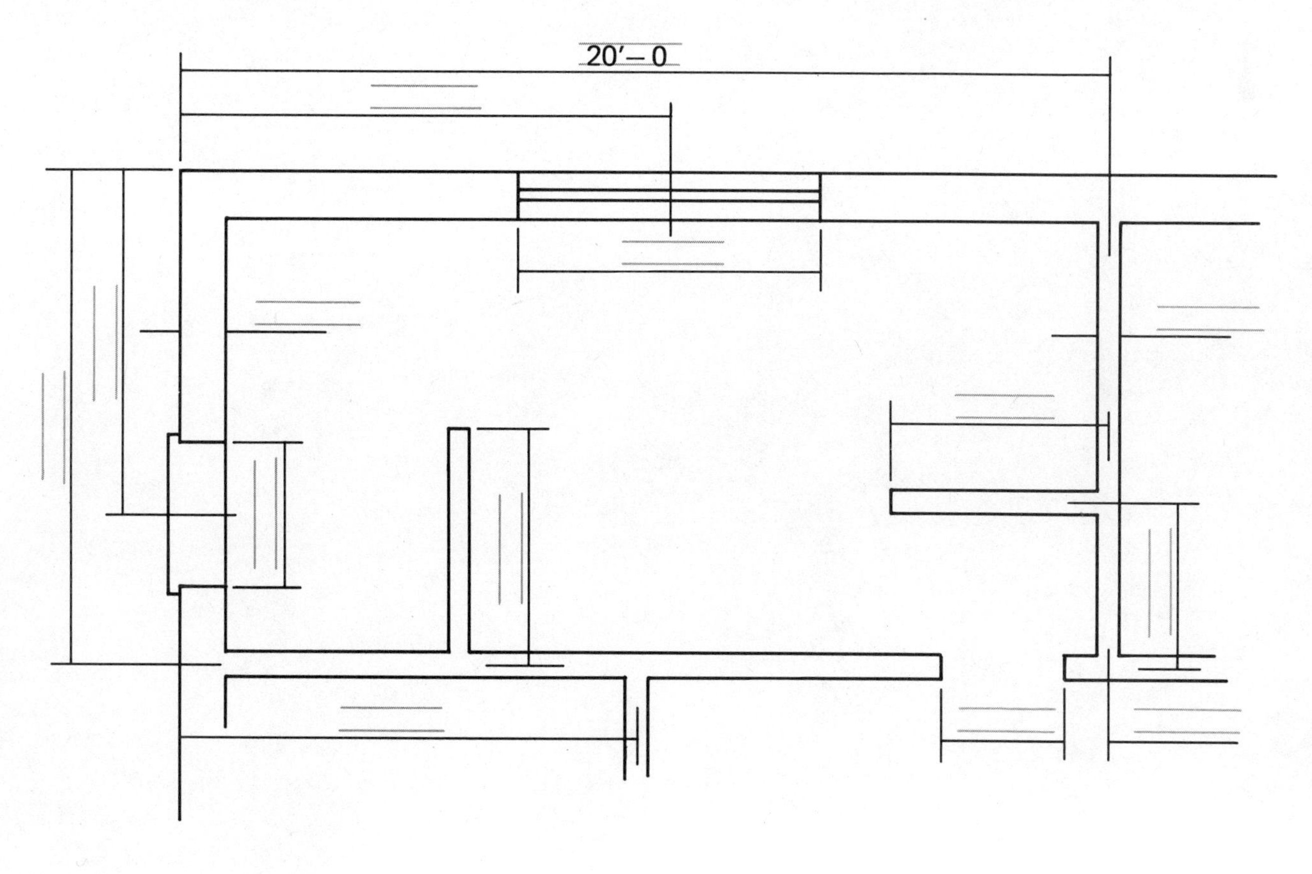

B

THE USE AND CARE OF DRAFTING EQUIPMENT	DRAWN BY	DATE	DWG NO.

DRAWN BY
DATE
DWG NO.

Determine the length of each lettered dimension on the scales below. Record each in the corresponding lettered space below. Work carefully within the guidelines. Add arrowheads.

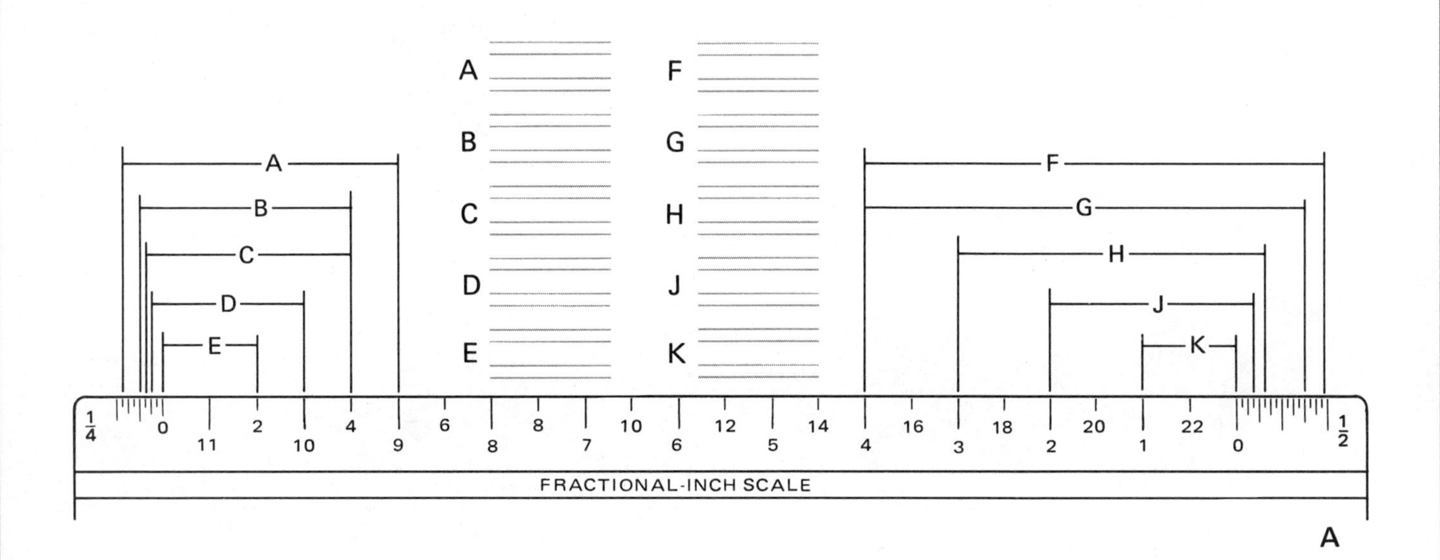

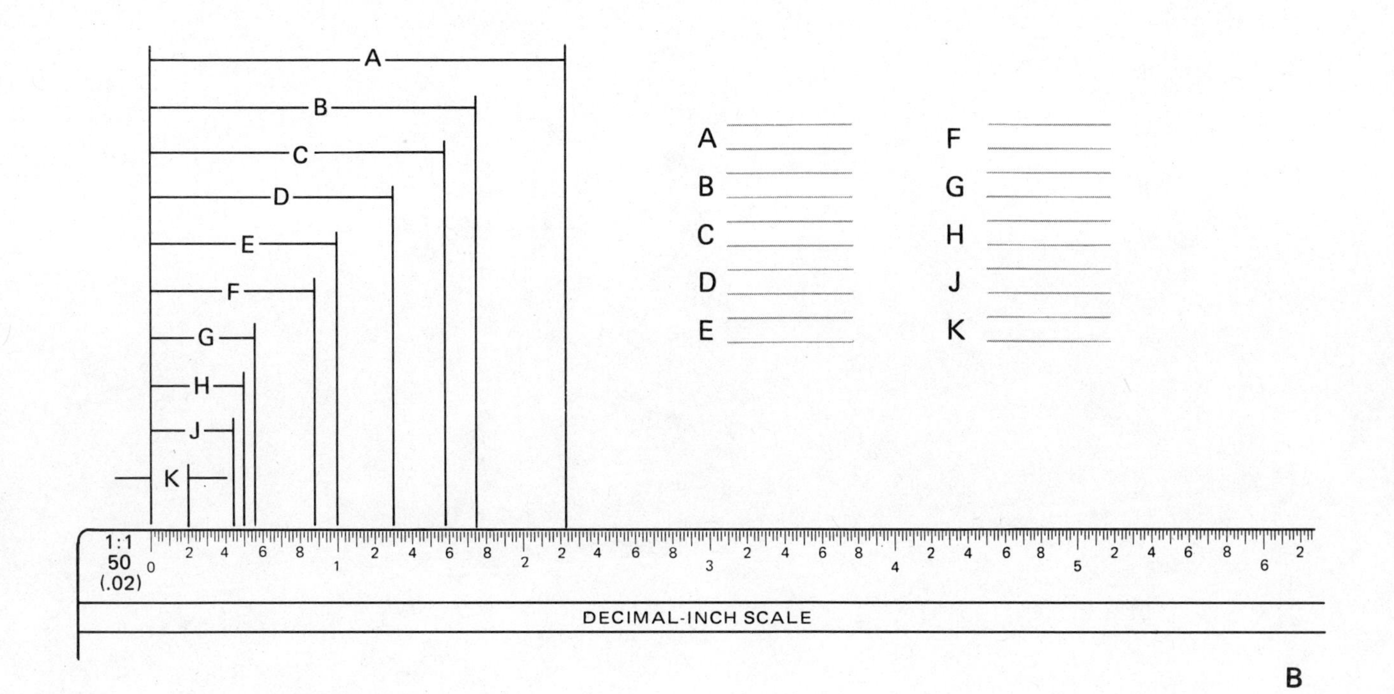

THE USE AND CARE OF DRAFTING EQUIPMENT

DRAWN BY	DATE	DWG NO.

DRAWN BY
DATE
DWG NO.

Determine the length of each of the lettered dimensions below using the civil engineer's 10 scale. Each division on the scale equals one-tenth inch. Dimensions that fall between divisions should be estimated (see example below) and given in two decimal places. Others should be given in one decimal place. Letter your answers in the corresponding spaces at the right. Add arrowheads.

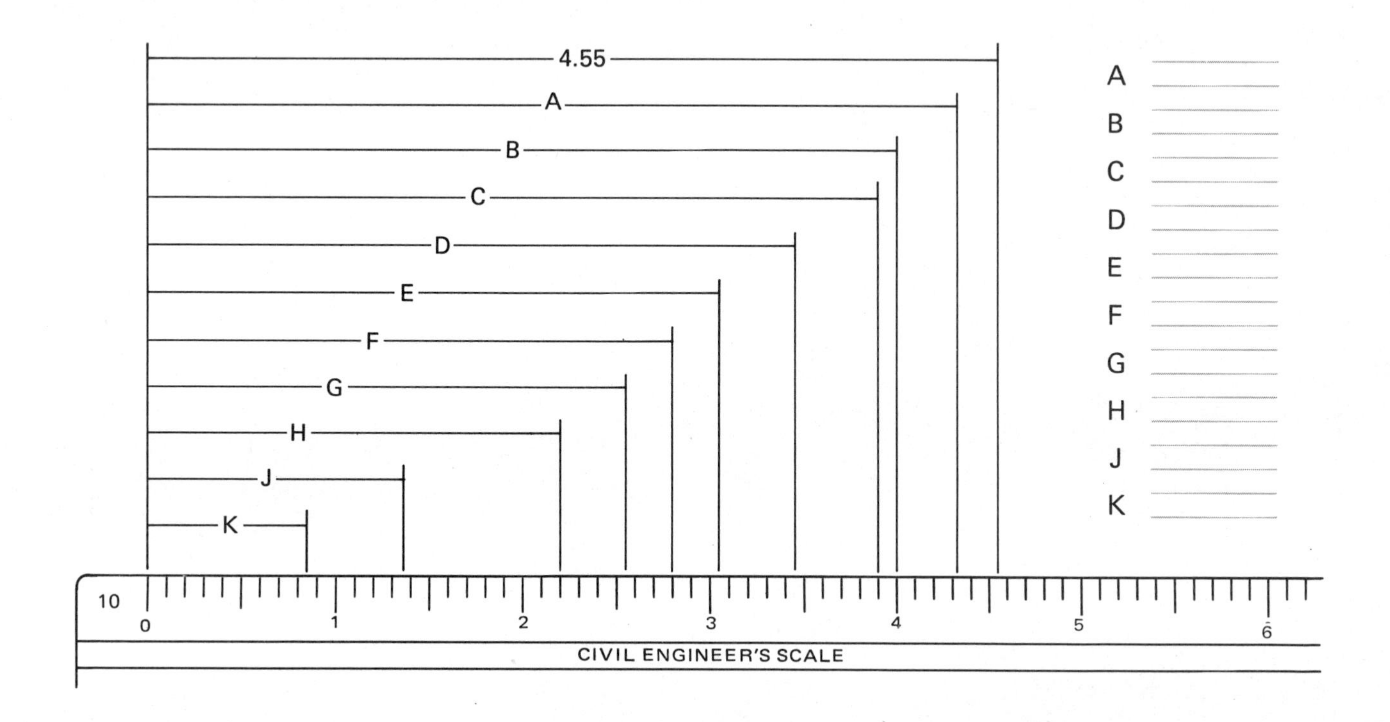

A

Determine the length of each of the lettered dimensions below using the civil engineer's 50 scale. Each division on the scale equals one-tenth inch. All dimensions should be given to one decimal place. Letter your answers in the corresponding spaces at the right. Add arrowheads.

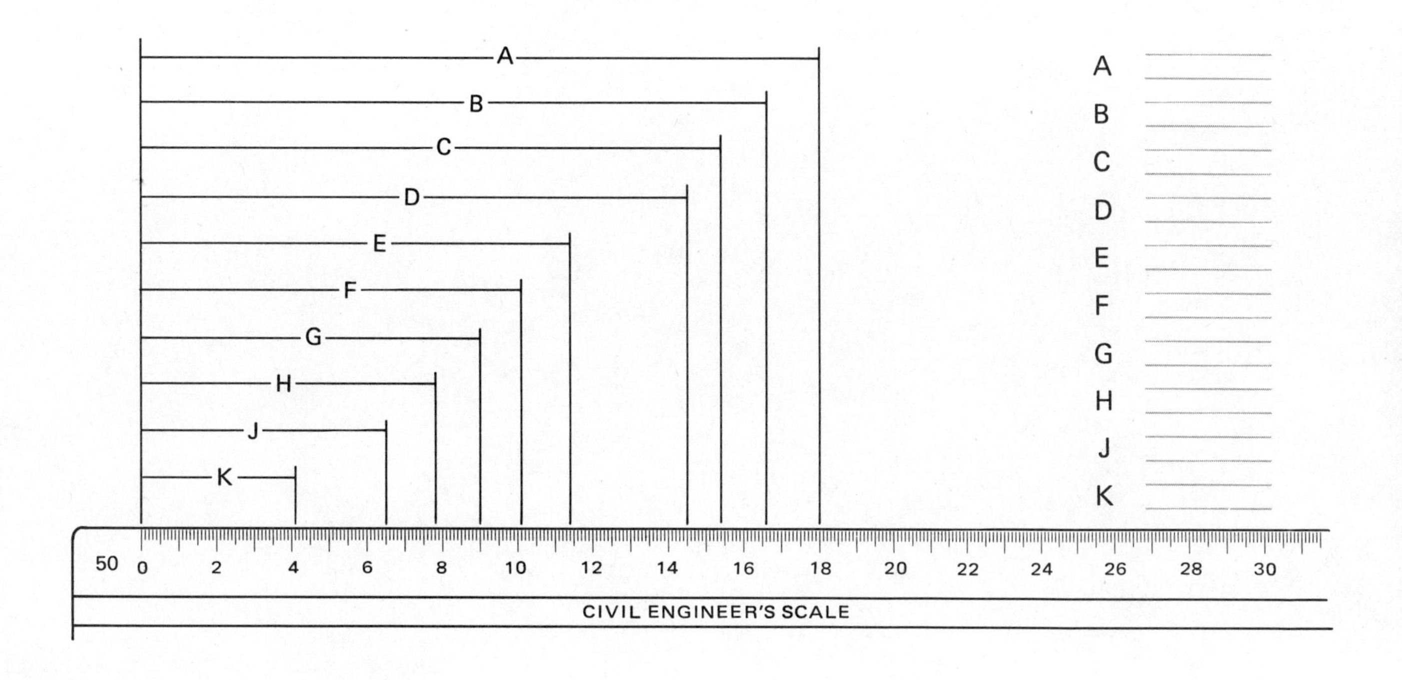

B

THE USE AND CARE OF DRAFTING EQUIPMENT	DRAWN BY	DATE	DWG NO.

DRAWN BY
DATE
DWG NO.

Determine the length of each of the lettered dimensions below using the full-size metric scale. Letter your answers in the corresponding spaces at the right. Specify sizes to the nearest half millimeter. Add arrowheads.

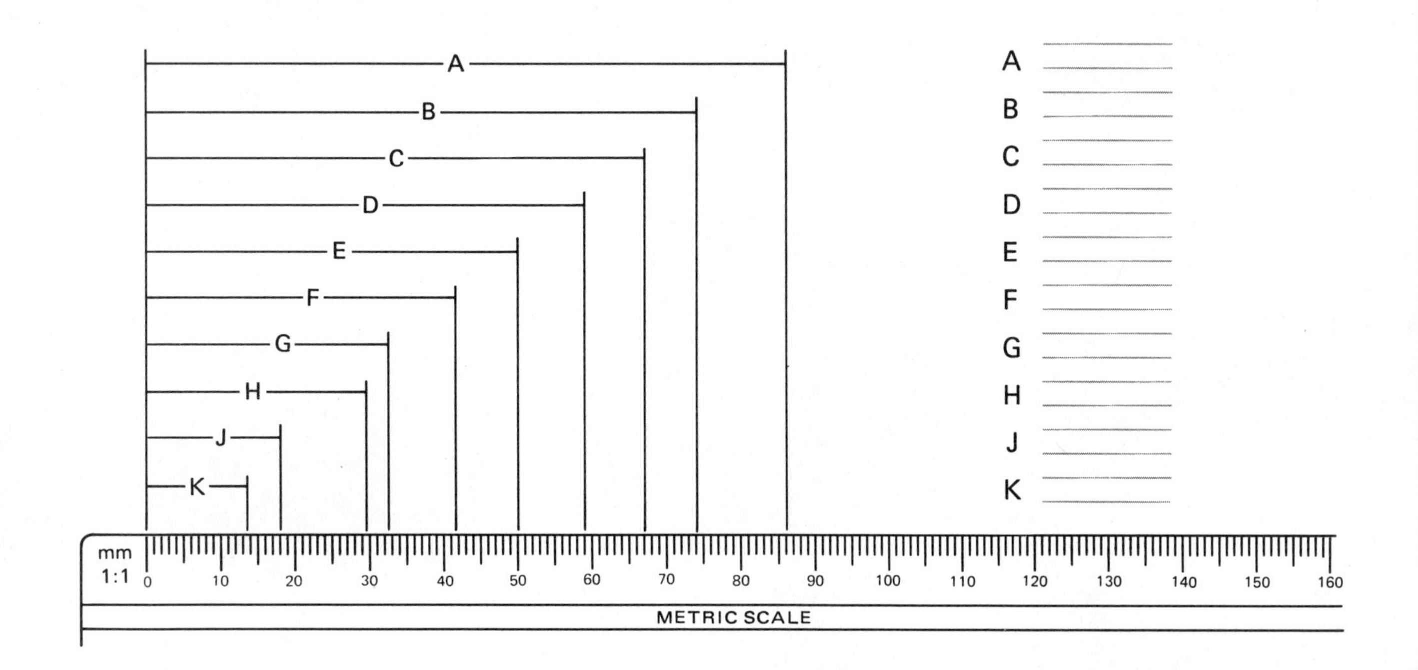

Determine the length of each of the lettered dimensions below on the metric 1:5 scale. Letter each in the corresponding spaces at the right. Specify sizes to the nearest millimeter. Add arrowheads.

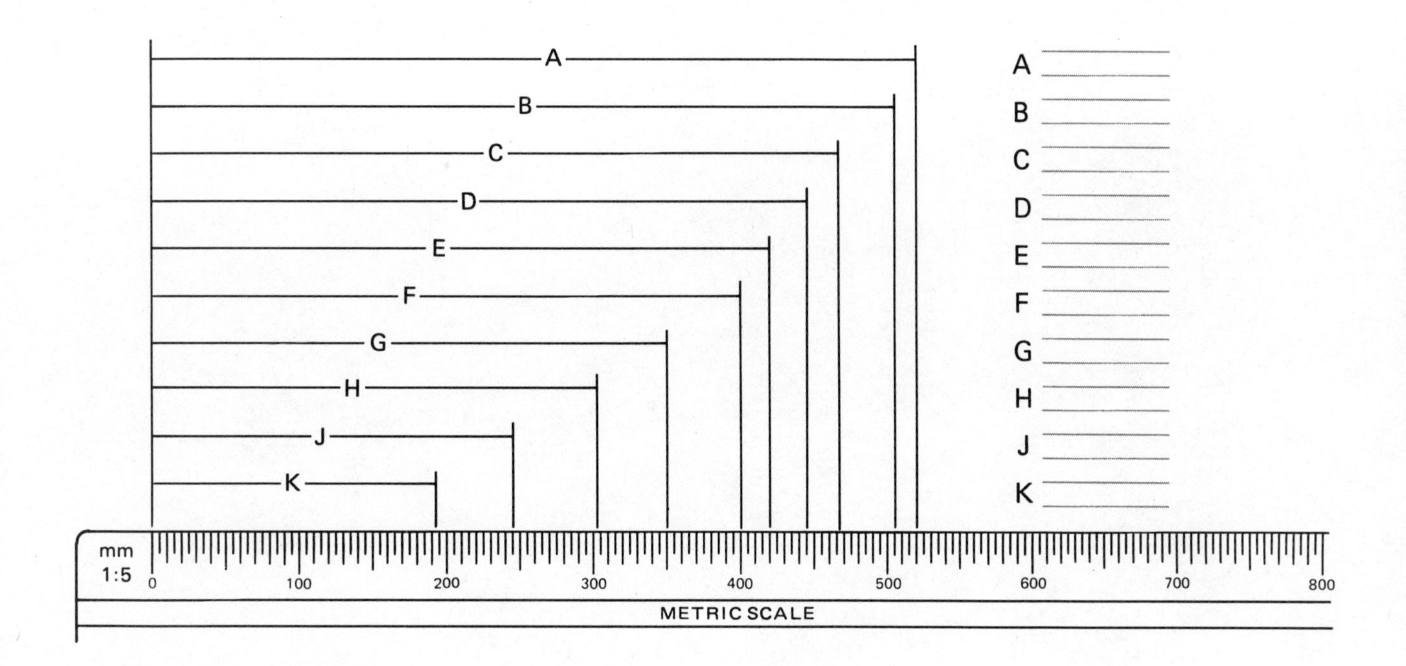

THE USE AND CARE OF DRAFTING EQUIPMENT	DRAWN BY	DATE	DWG NO.

DRAWN BY
DATE
DWG NO.

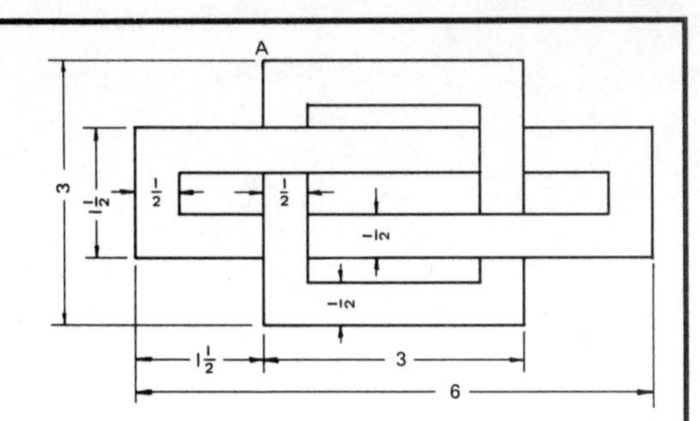

Beginning at point A, make a drawing of the *Inlay*.
Scale: Full size. Do not add dimensions unless instructed to do so.

A

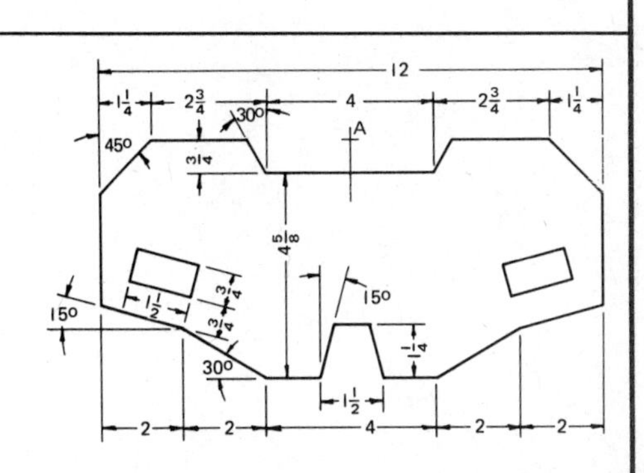

Beginning at point A, make a drawing of the *Shearing Blank*.
Scale: Half size. Do not add dimensions unless instructed to do so.

THE USE AND CARE OF DRAFTING EQUIPMENT	DRAWN BY	DATE	DWG NO.

DRAWN BY
DATE
DWG NO.

Beginning at point A, make a drawing of the *Template.* Scale: Full size. Do not add dimensions unless instructed to do so.

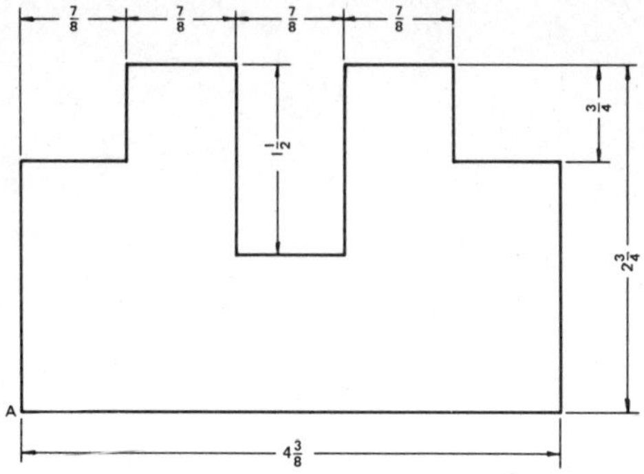

A

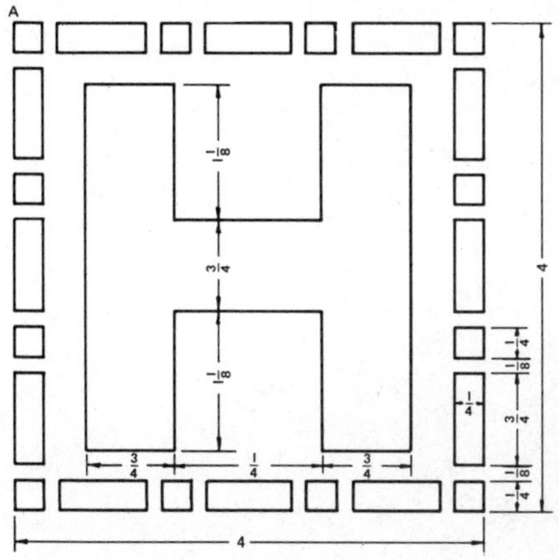

A

Beginning at point A, make a drawing of the *Stencil*. Scale: Full size. Do not add dimensions unless instructed to do so.

THE USE AND CARE OF DRAFTING EQUIPMENT	DRAWN BY	DATE	DWG NO.

DRAWN BY
DATE
DWG NO.

The left half of a template is shown below. Draw the right half so that the complete pattern will be symmetrical about the centerline. Measure the distances from the left half.

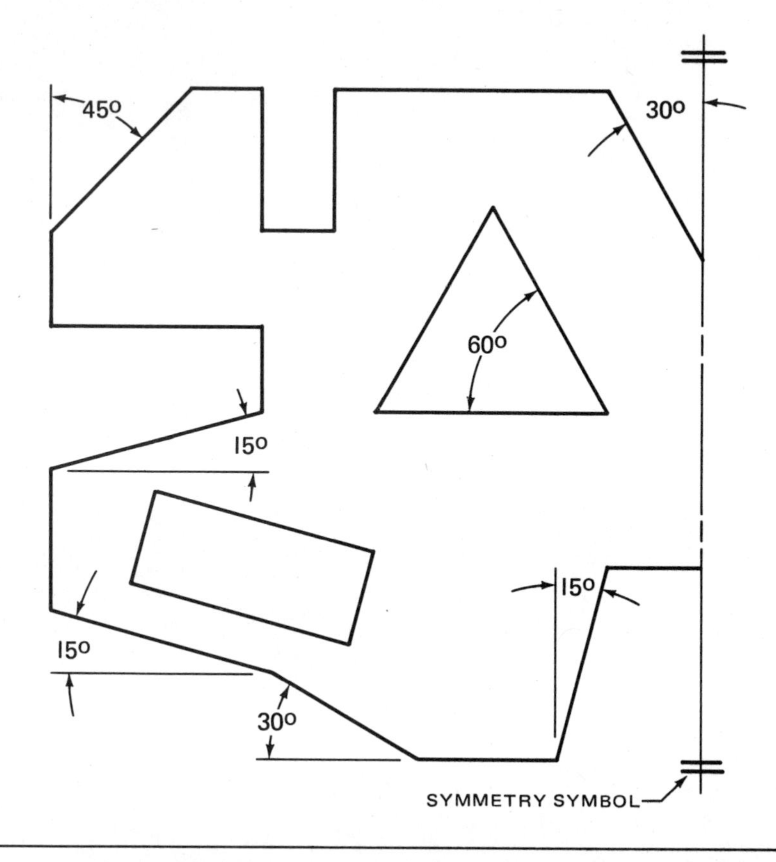

Complete the left half of the *Gasket* which is symmetrical about the centerline. Mark all points of tangency. Show all construction lines.

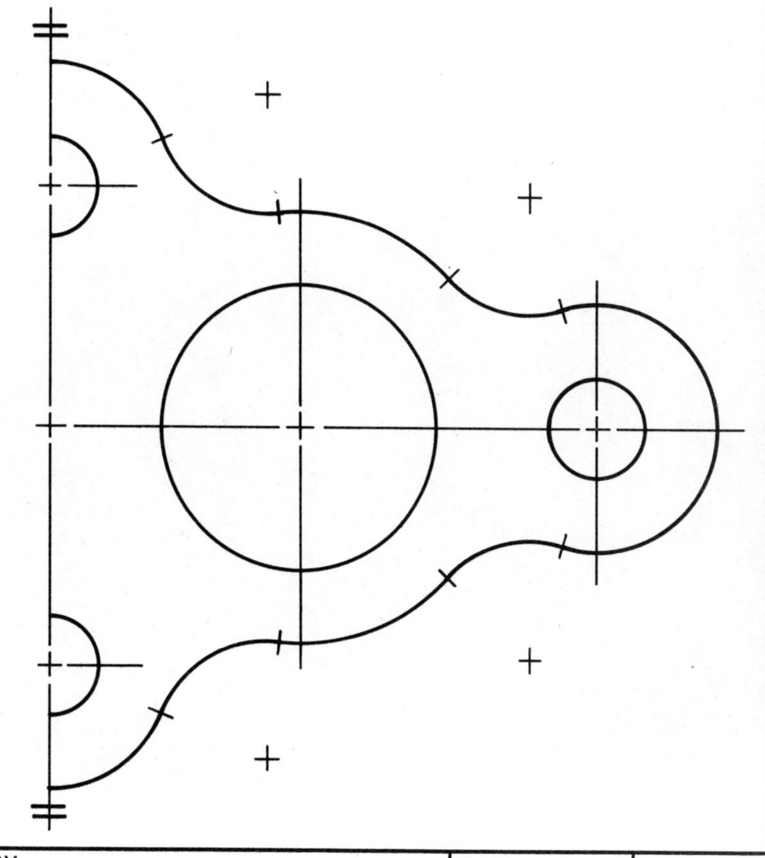

THE USE AND CARE OF DRAFTING EQUIPMENT	DRAWN BY	DATE	DWG NO.

DRAWN BY
DATE
DWG NO.

Beginning at point A, make a drawing of the *Adjustable Sector.* Scale: 3/4" = 1". Do not dimension unless instructed to do so. Small arcs may be drawn with a circle template. Work accurately and be sure to locate all points of tangency.

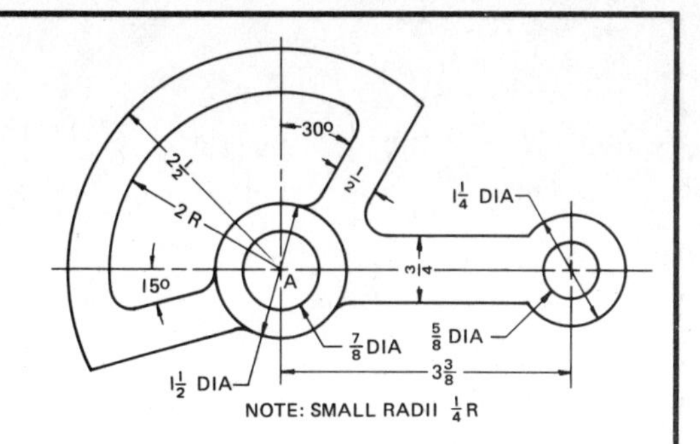

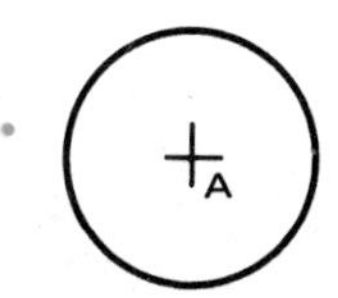

A

Beginning at point A, make a drawing of the *Cushioning Base.* Scale: 1" = 1'. Do not dimension unless instructed to do so. Work accurately and be sure to locate all points of tangency.

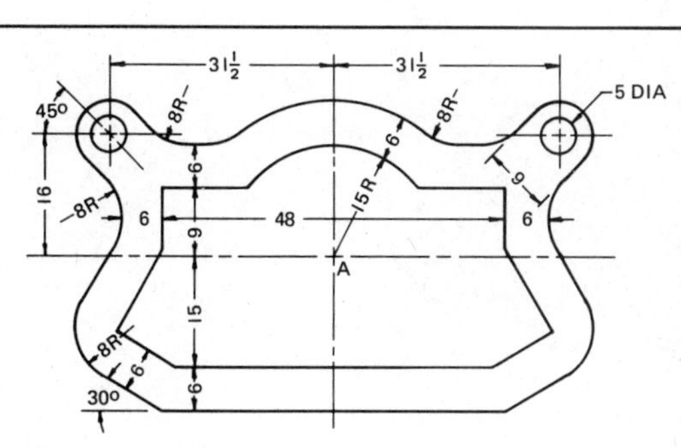

B

THE USE AND CARE OF DRAFTING EQUIPMENT	DRAWN BY	DATE	DWG NO.

DRAWN BY
DATE
DWG NO.

1. Begin at point A and complete the eight-sided figure shown at the right. Work clockwise.
2. Divide line AB into five equal parts.
3. Bisect angle GHA. Measure and record the resulting angles.
4. Divide line CD into four equal parts. How long is each part?
5. Bisect angle DEF. Measure and record the resulting angles.
6. Draw a line parallel to line AH beginning at point P and ending at line AB.
7. Bisect line BC.
8. Begin at point X and draw the *Fast-Forward Symbol* full size. Do not dimension.

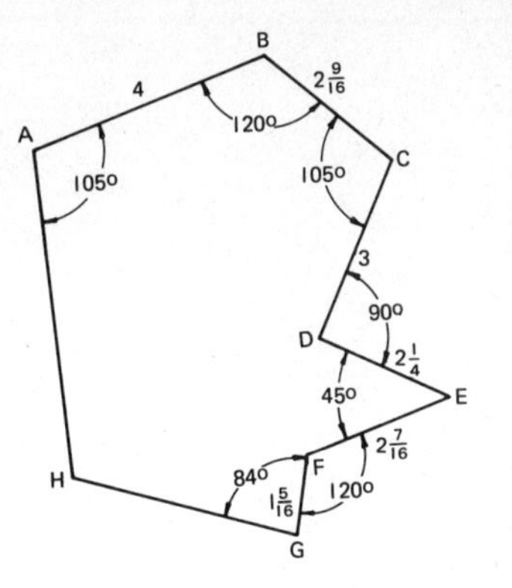

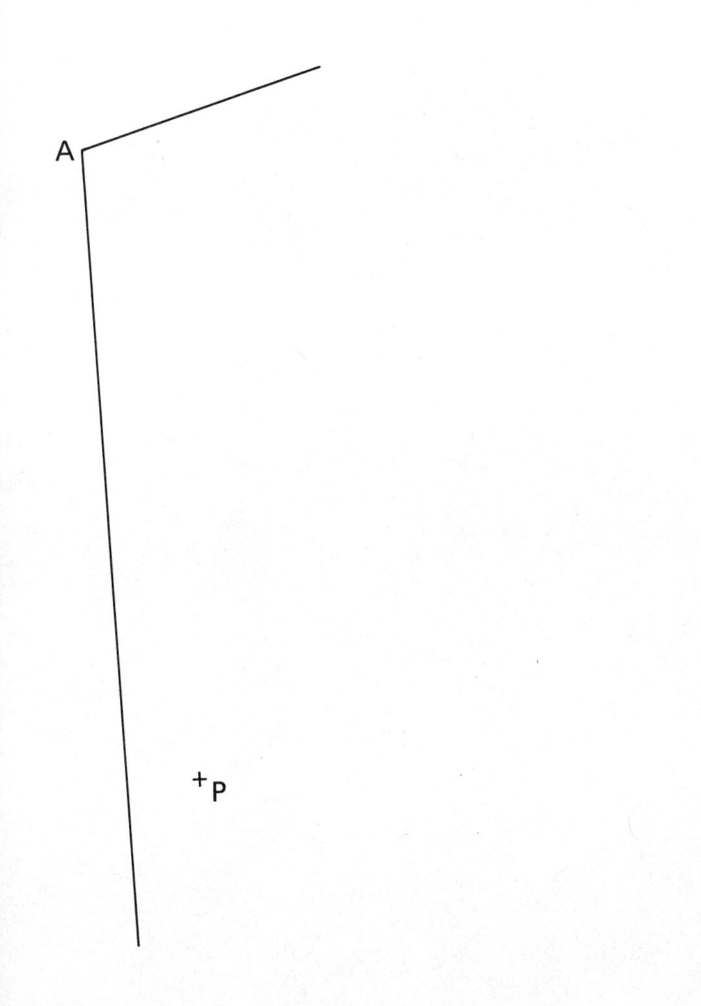

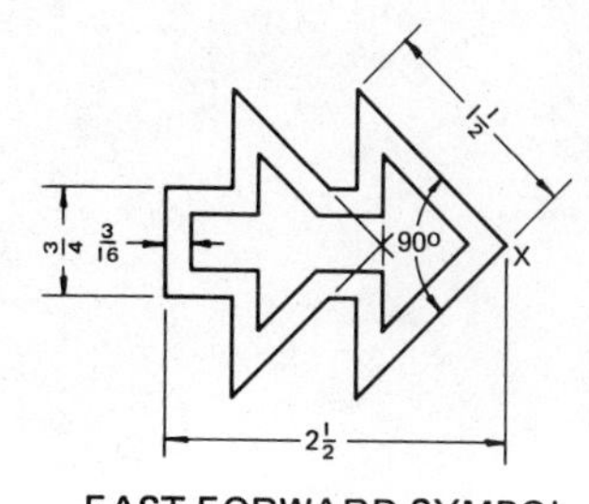

FAST-FORWARD SYMBOL

GEOMETRY FOR TECHNICAL DRAWING	DRAWN BY	DATE	DWG NO.

	DRAWN BY	DATE	DWG NO.

Draw lines tangent to the circle at points A through H to form an octagon. Show all construction lines.

A
H
B
G
C
F
D
E

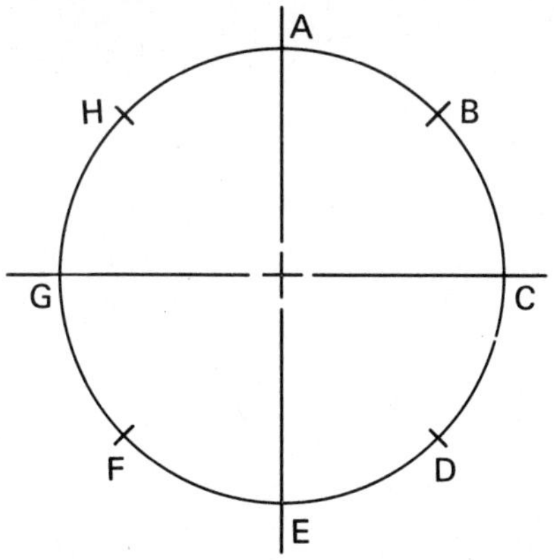

A

From point A, draw lines tangent to the circle at points B and C. From point D, draw lines tangent to the circle at points E and F. Show all construction lines.

E
B
D
A
C
F

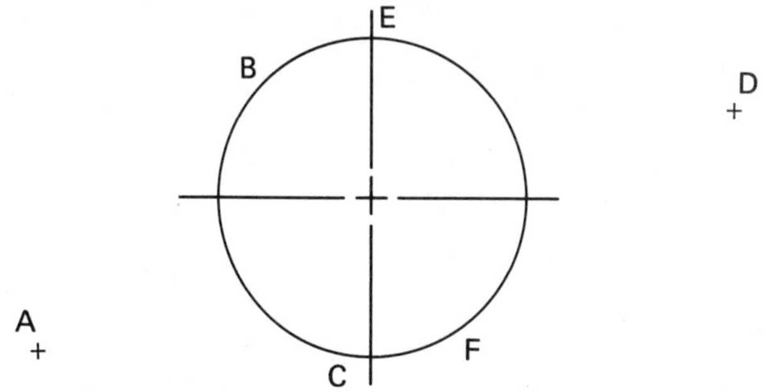

B

Construct a circle through points A, B, and C. Show all construction lines.

B
A
C

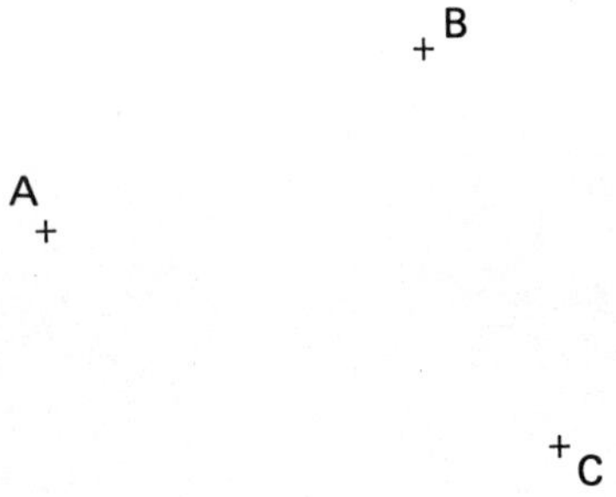

C

Construct a hexagon having a distance of 45mm across the flats. Show all construction lines.

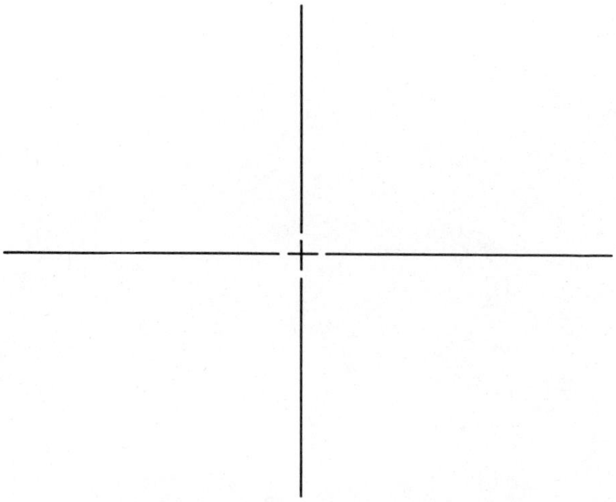

D

Construct a hexagon having line AB as the distance across the corners. Show all construction lines.

A B

E

Construct a pentagon in the circle below. Add lines to construct a star inside the pentagon. Show all construction lines.

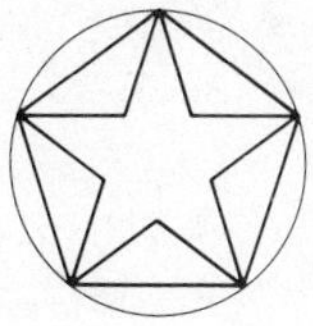

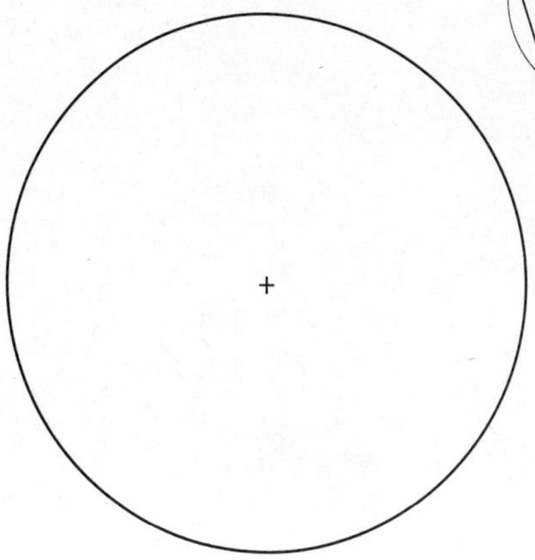

F

	DRAWN BY	DATE	DWG NO.

	DRAWN BY	DATE	DWG NO.

Complete the front view. Mark all points of tangency with a short dash. Do not dimension. Use a compass for all circles and arcs. Draw circles and arcs before drawing straight lines.

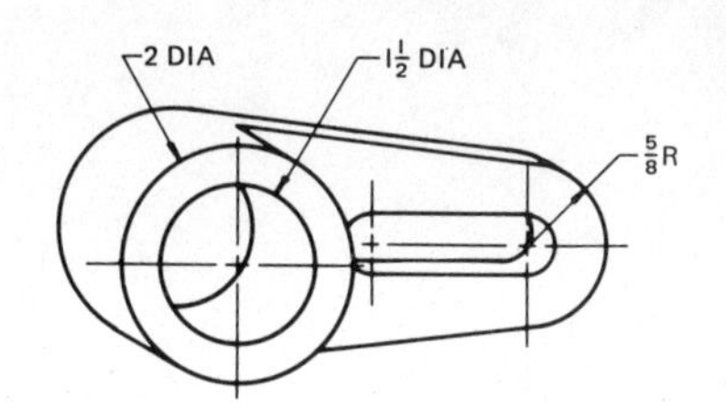

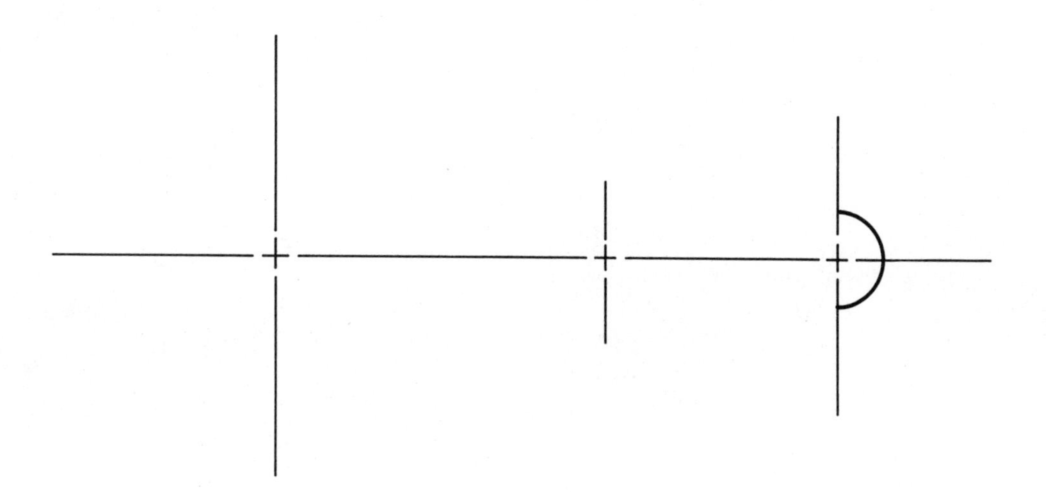

A

Complete the front view of the *Brace.* Carefully locate and mark all points of tangency with a short dash. Do not dimension. Use a compass for all circles and arcs. Draw circles and arcs before drawing straight lines.

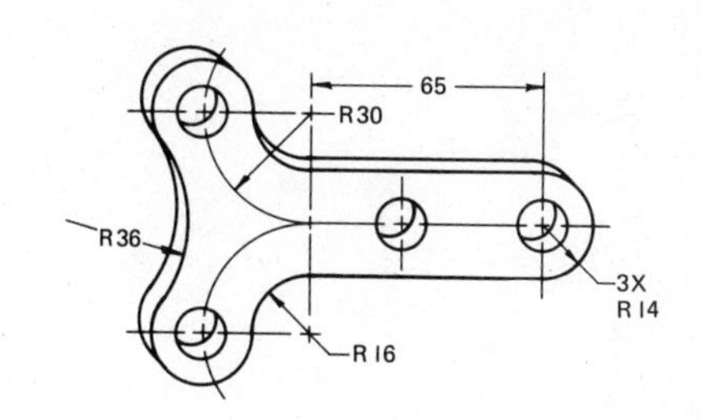

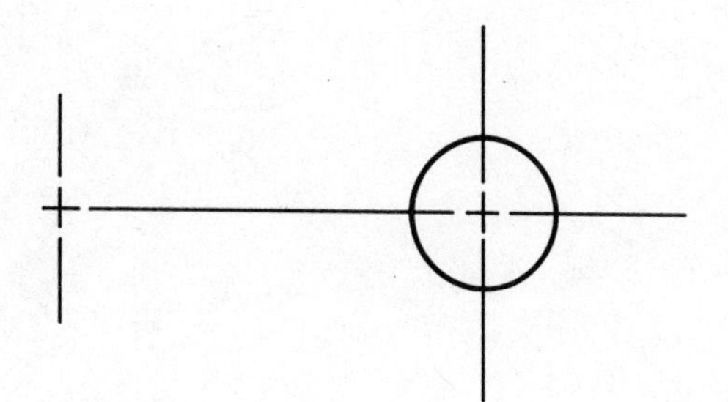

METRIC

B

GEOMETRY FOR TECHNICAL DRAWING	DRAWN BY	DATE	DWG NO.

	DRAWN BY	DATE	DWG NO.

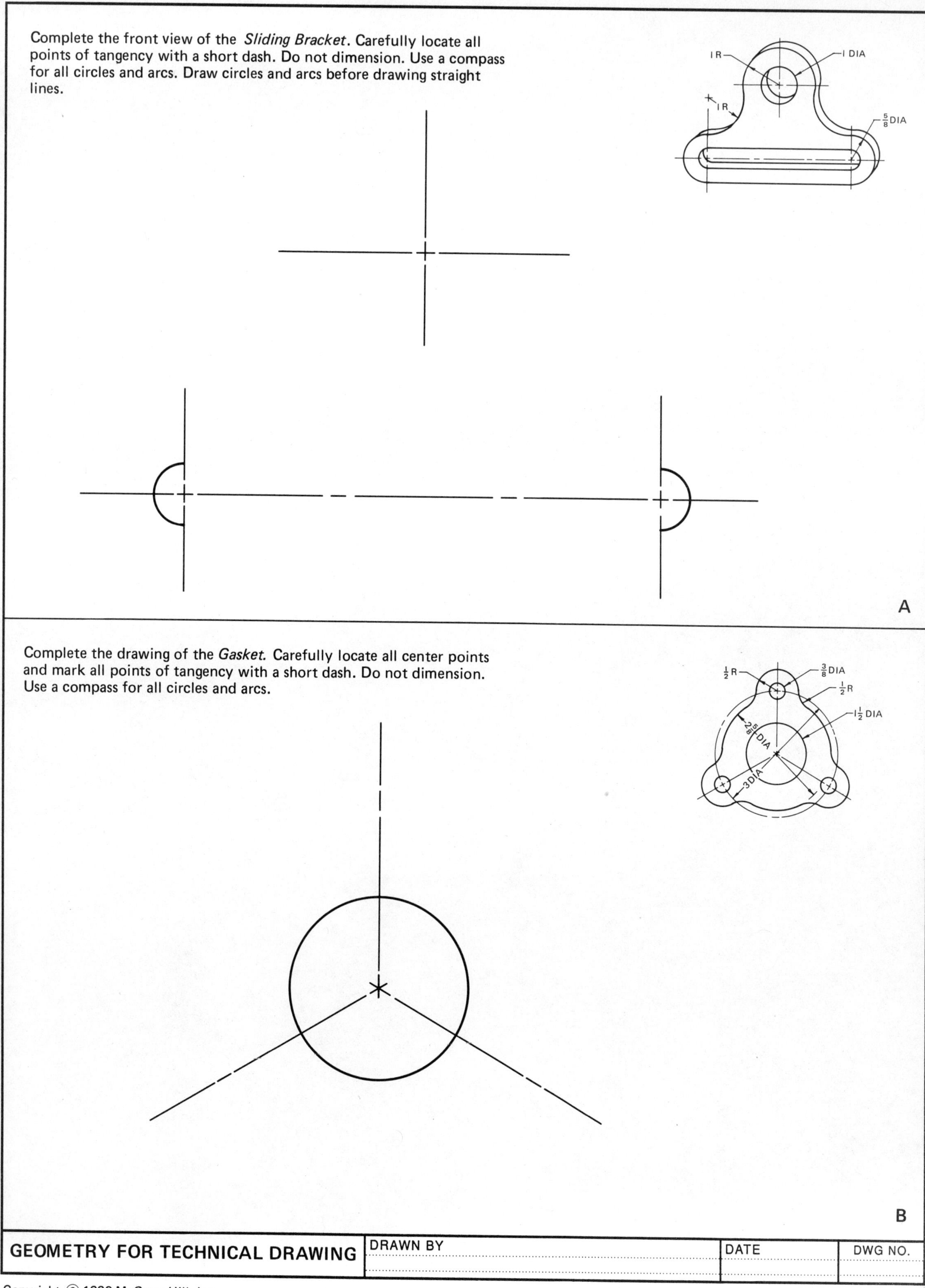

Complete the front view of the *Sliding Bracket*. Carefully locate all points of tangency with a short dash. Do not dimension. Use a compass for all circles and arcs. Draw circles and arcs before drawing straight lines.

Complete the drawing of the *Gasket.* Carefully locate all center points and mark all points of tangency with a short dash. Do not dimension. Use a compass for all circles and arcs.

GEOMETRY FOR TECHNICAL DRAWING	DRAWN BY	DATE	DWG NO.

	DRAWN BY	DATE	DWG NO.

Draw the front view of the *Arch* on the centerlines given. Use only a compass to draw all arcs. Locate and mark points of tangency. Carefully blend arcs with other arcs and arcs with straight lines. Do not dimension.

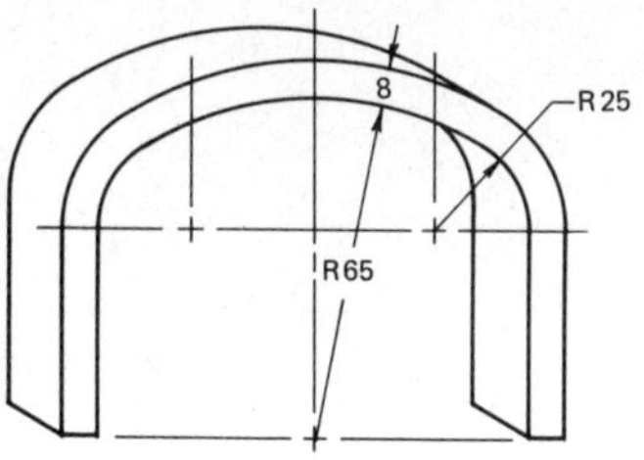

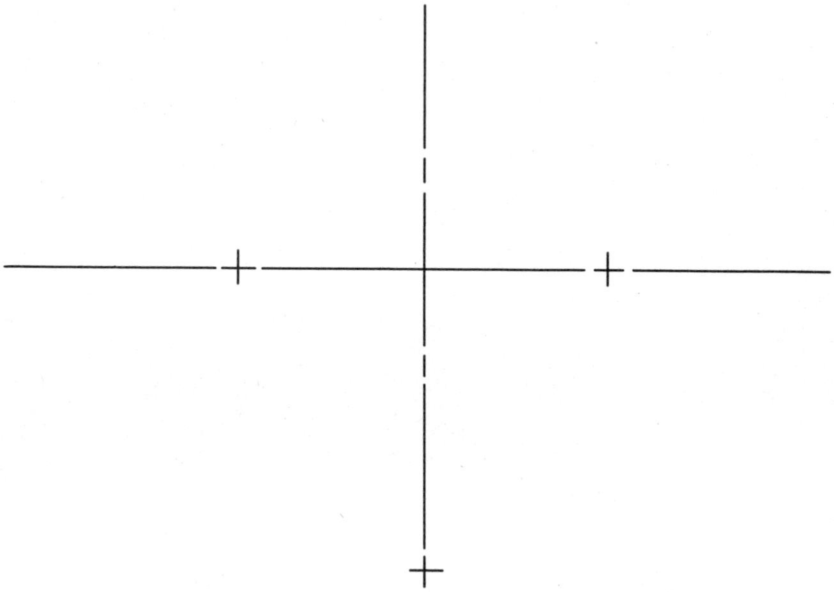

METRIC

A

Draw the front view of the *Pivot Link* on the centerlines given. Use only a compass to locate required center points and to draw arcs. Locate and mark points of tangency. Do not dimension.

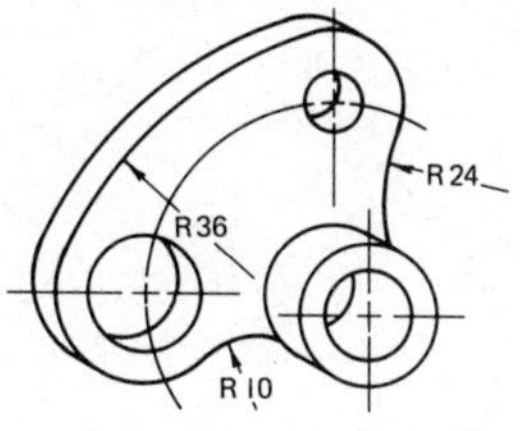

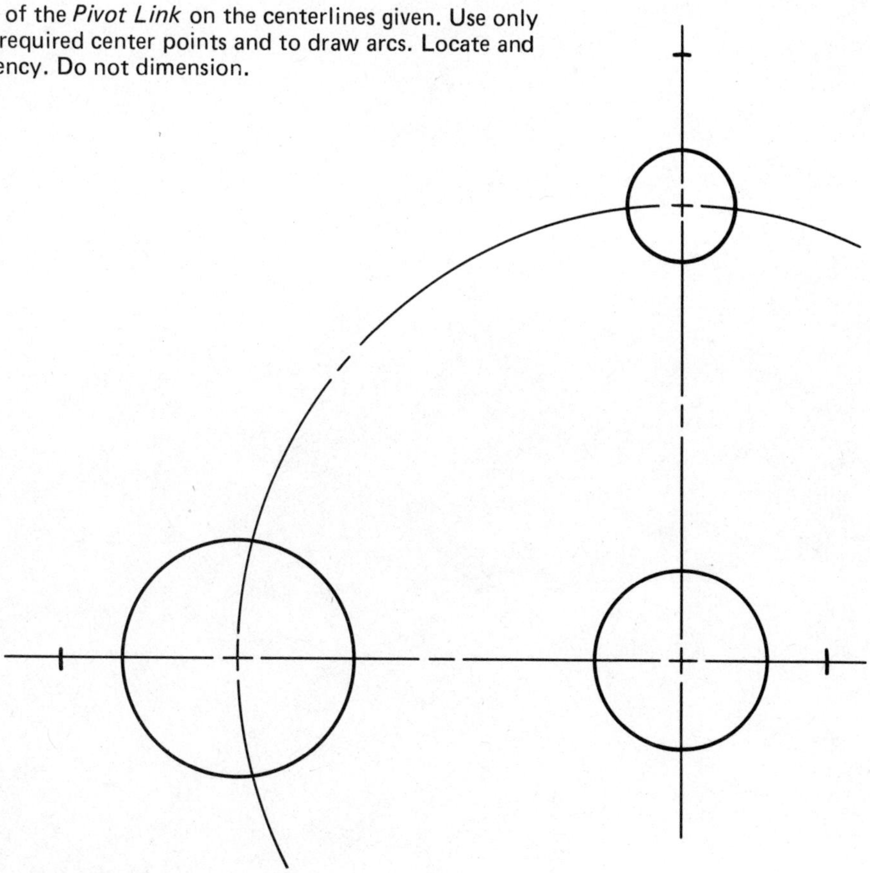

METRIC

B

GEOMETRY FOR TECHNICAL DRAWING	DRAWN BY	DATE	DWG NO.

DRAWN BY
DATE
DWG NO.

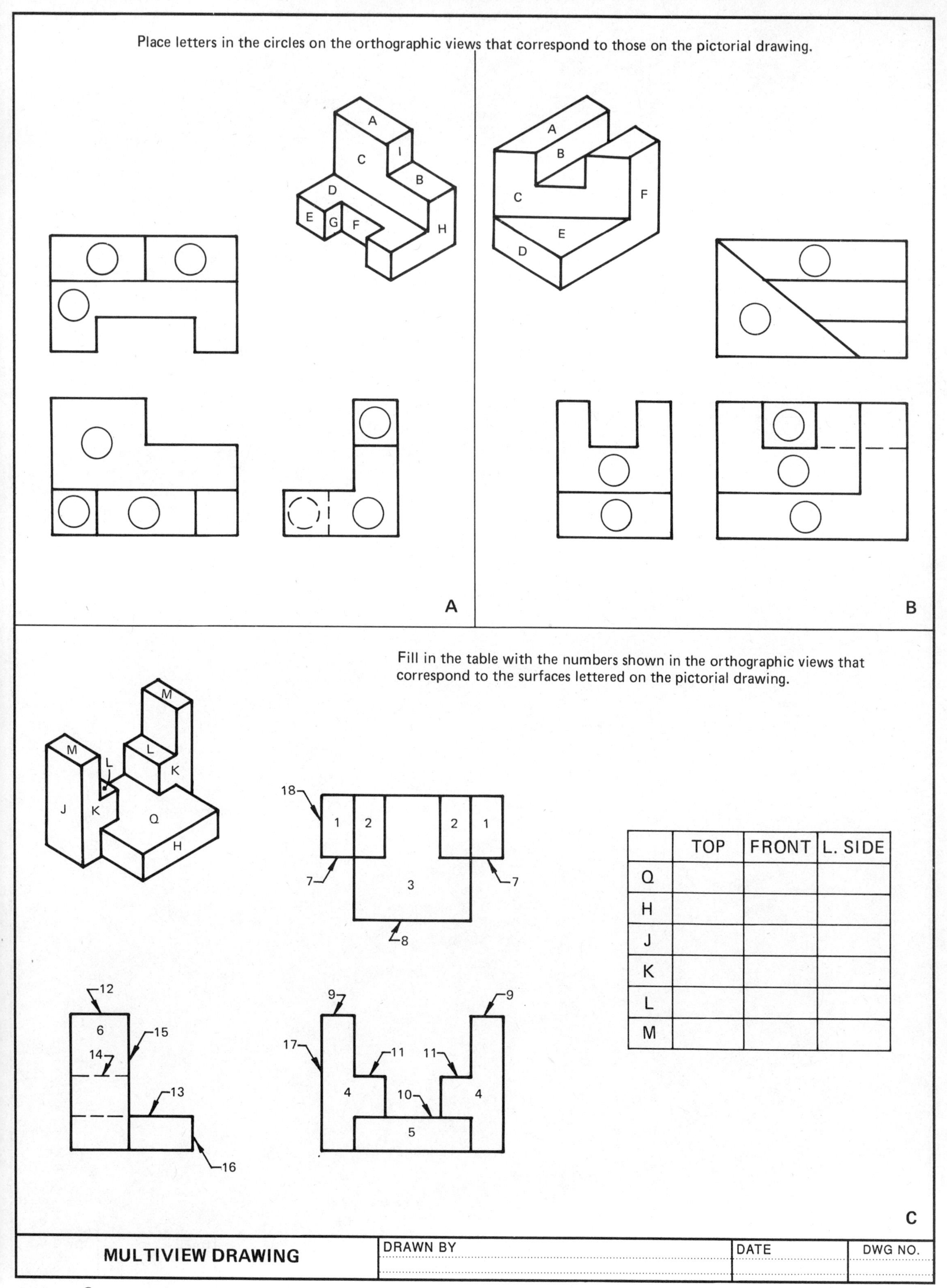

	TOP	FRONT	L. SIDE
Q			
H			
J			
K			
L			
M			

DRAWN BY
DATE
DWG NO.

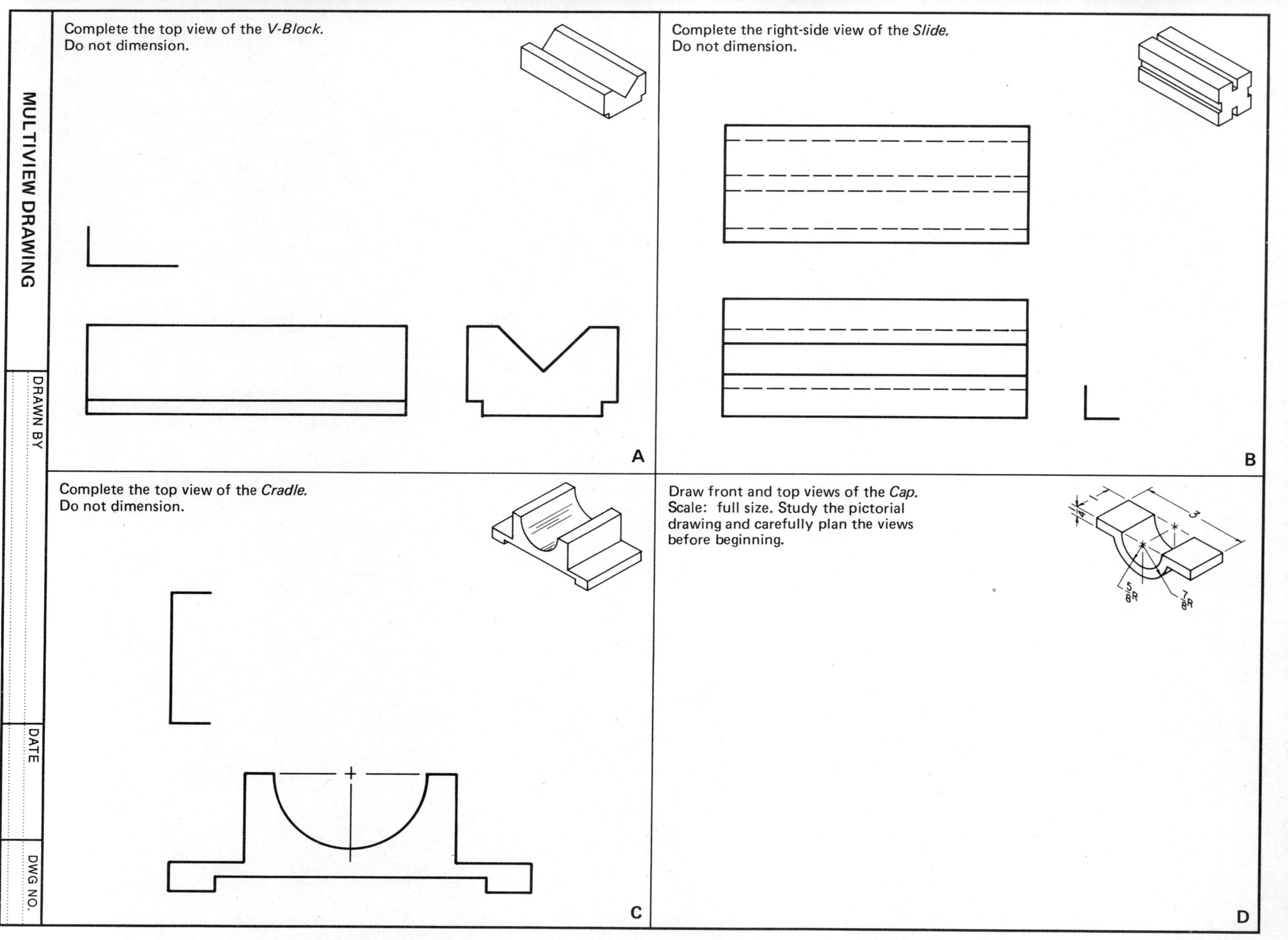

Complete the top view of the *V-Block.*
Do not dimension.
A
Complete the right-side view of the *Slide.*
Do not dimension.
B
Complete the top view of the *Cradle.*
Do not dimension.
C
Draw front and top views of the *Cap.*
Scale: full size. Study the pictorial
drawing and carefully plan the views
before beginning.
3
$\frac{5}{8}$R
$\frac{7}{8}$R
D
MULTIVIEW DRAWING
DRAWN BY
DATE
DWG NO.

DRAWN BY
DATE
DWG NO.

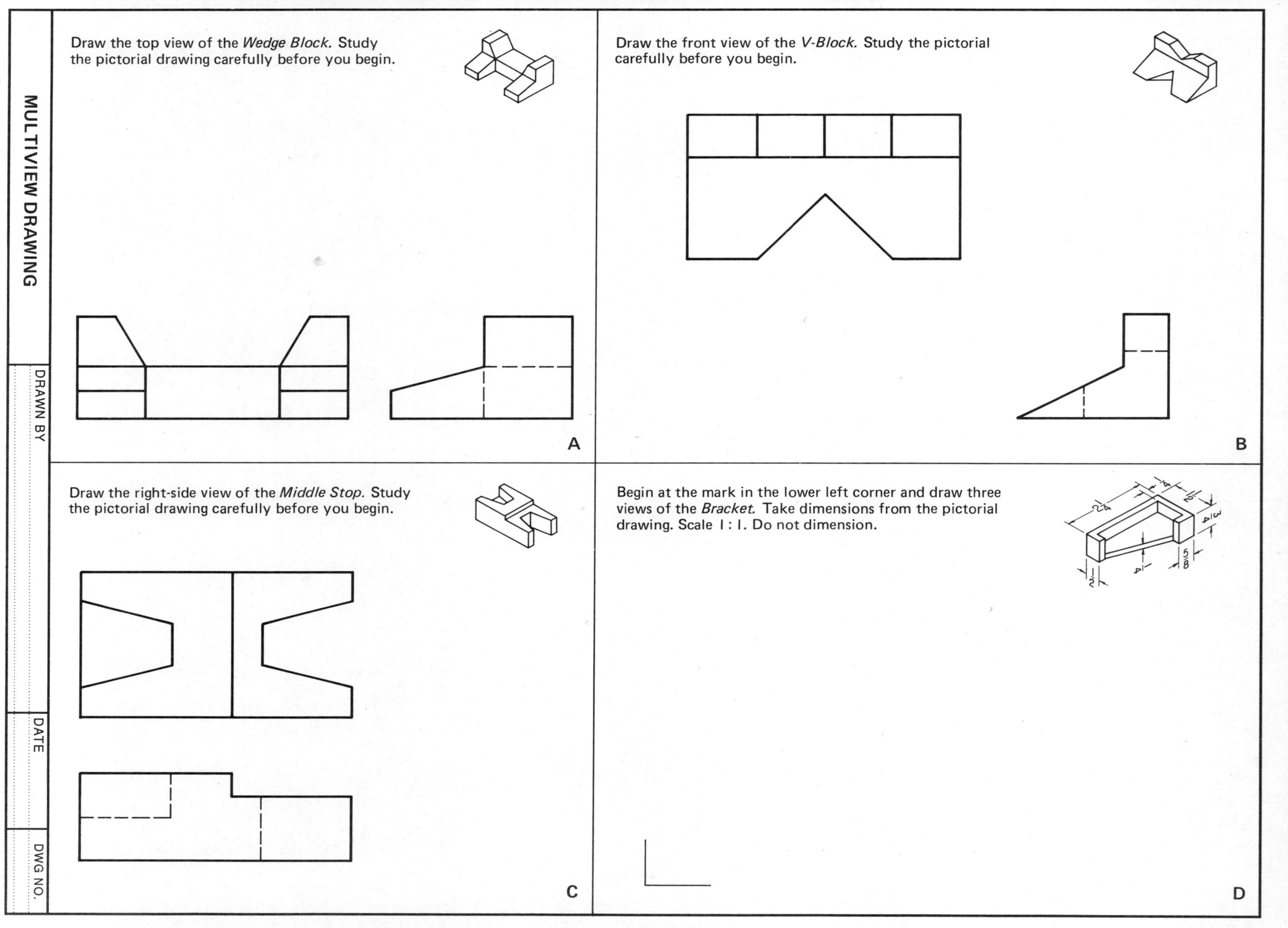
MULTIVIEW DRAWING
Draw the top view of the *Wedge Block.* Study the pictorial drawing carefully before you begin.
A
Draw the front view of the *V-Block.* Study the pictorial carefully before you begin.
B
Draw the right-side view of the *Middle Stop.* Study the pictorial drawing carefully before you begin.
C
Begin at the mark in the lower left corner and draw three views of the *Bracket.* Take dimensions from the pictorial drawing. Scale 1 : 1. Do not dimension.
D
DRAWN BY
DATE
DWG NO.

DRAWN BY
DATE
DWG NO.

Draw the top view of the *Support Guide.* Do not erase construction/projection lines. Do not dimension.

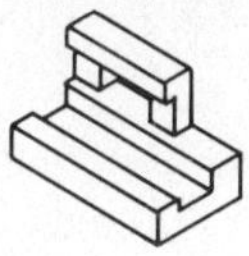

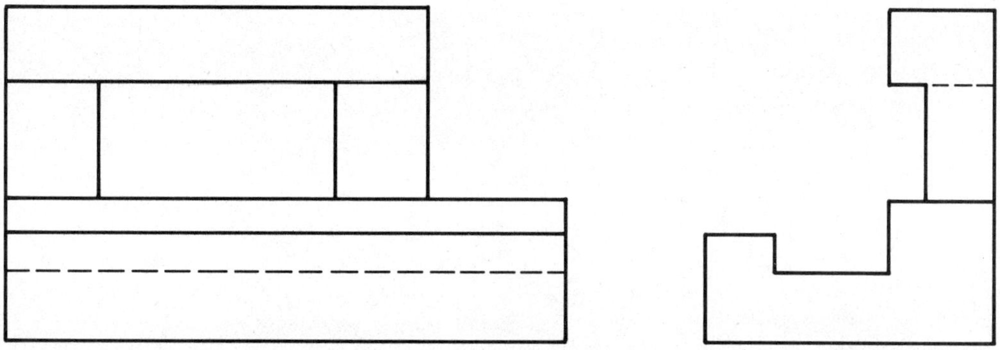

A

Draw the front view of the *Angle Bracket.* Do not erase construction/projection lines. Do not dimension.

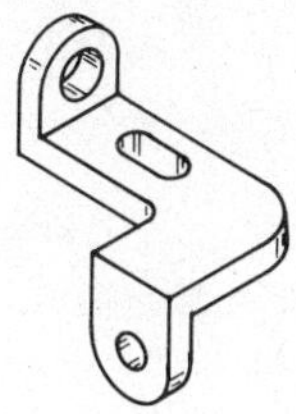

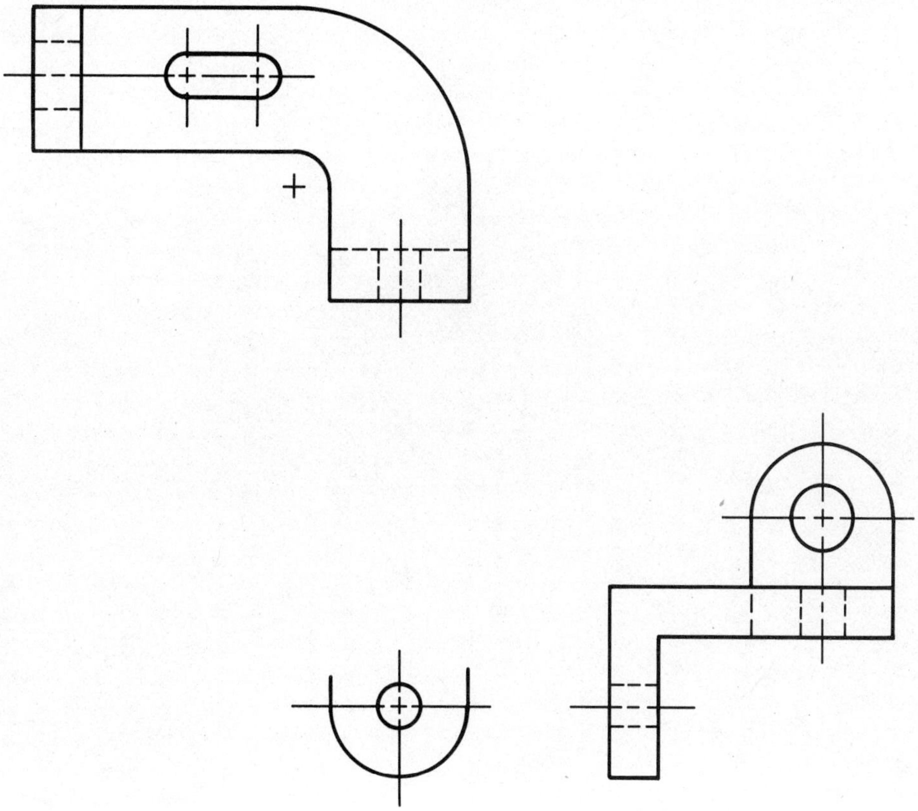

B

MULTIVIEW DRAWING	DRAWN BY	DATE	DWG NO.

DRAWN BY
DATE
DWG NO.

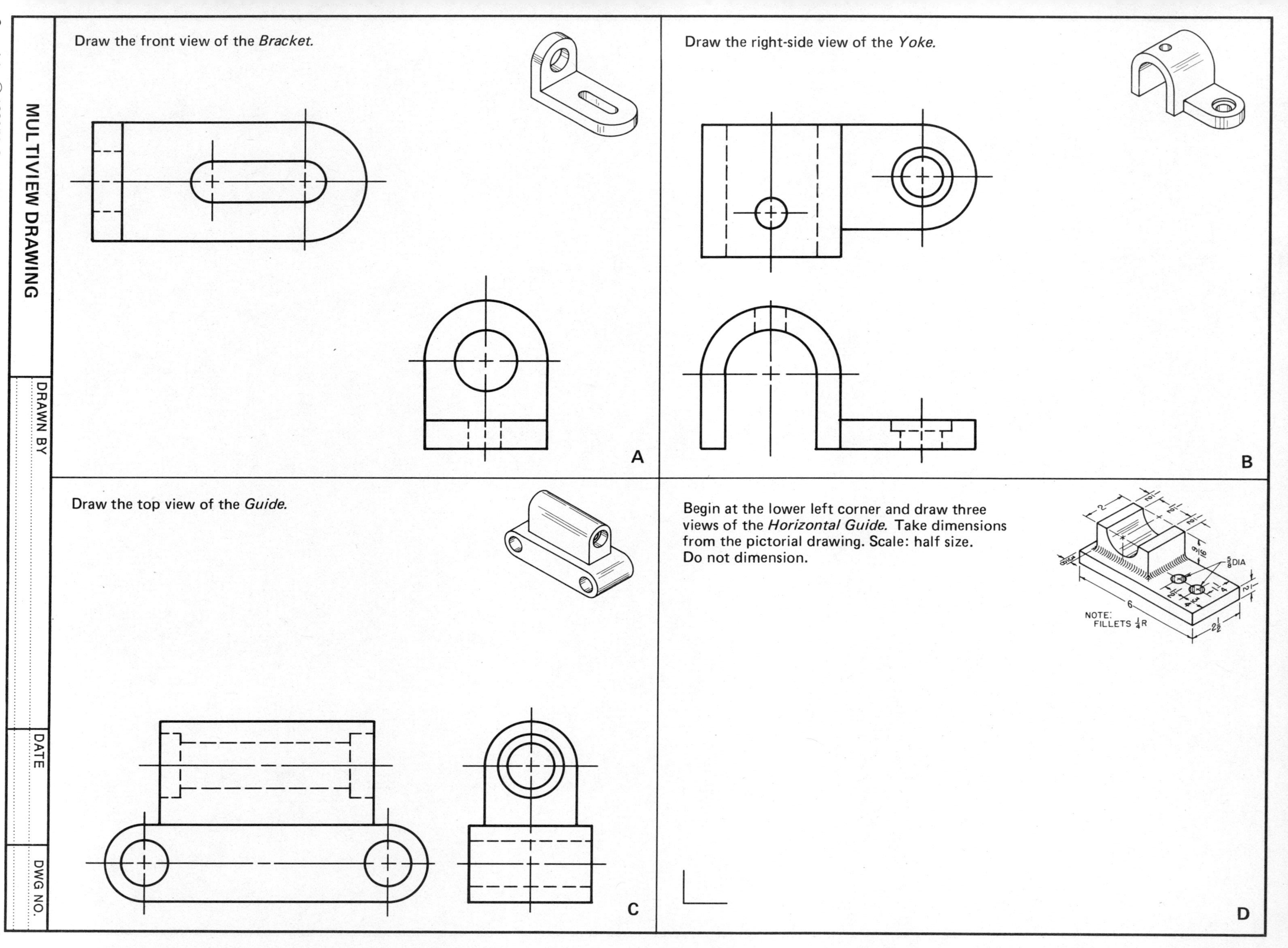
Draw the front view of the *Bracket*.
A
Draw the right-side view of the *Yoke*.
B
Draw the top view of the *Guide*.
C
Begin at the lower left corner and draw three views of the *Horizontal Guide*. Take dimensions from the pictorial drawing. Scale: half size. Do not dimension.
NOTE: FILLETS 1/4 R
5/8 DIA
D
MULTIVIEW DRAWING
DRAWN BY
DATE
DWG NO.

DRAWN BY
DATE
DWG NO.

Draw three views of the *Slide.* Scale: half size. Do not dimension.

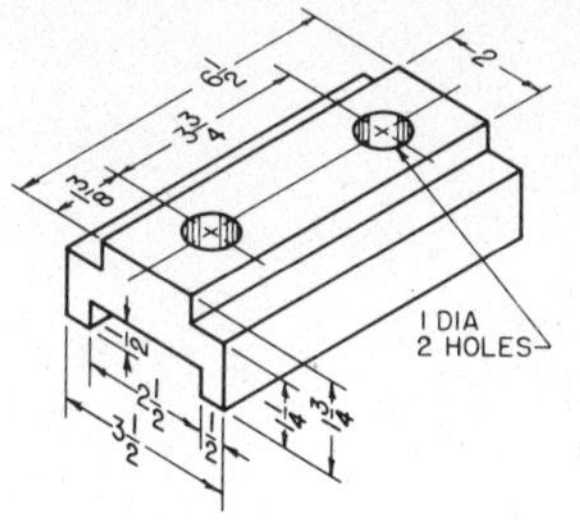

A

Draw three views of the *Vise Base.* Scale: half size. Do not dimension.

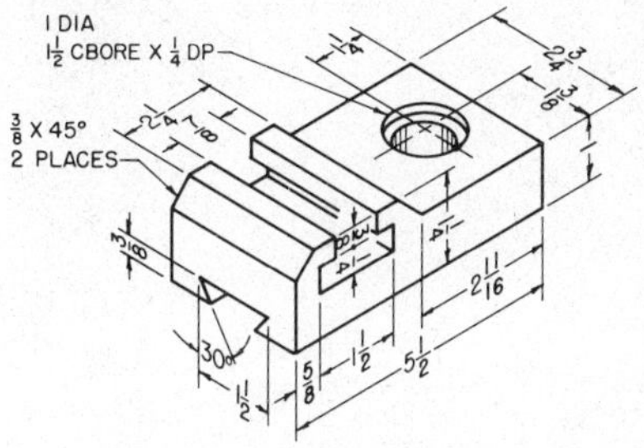

B

MULTIVIEW DRAWING	DRAWN BY	DATE	DWG NO.

DRAWN BY
DATE
DWG NO.

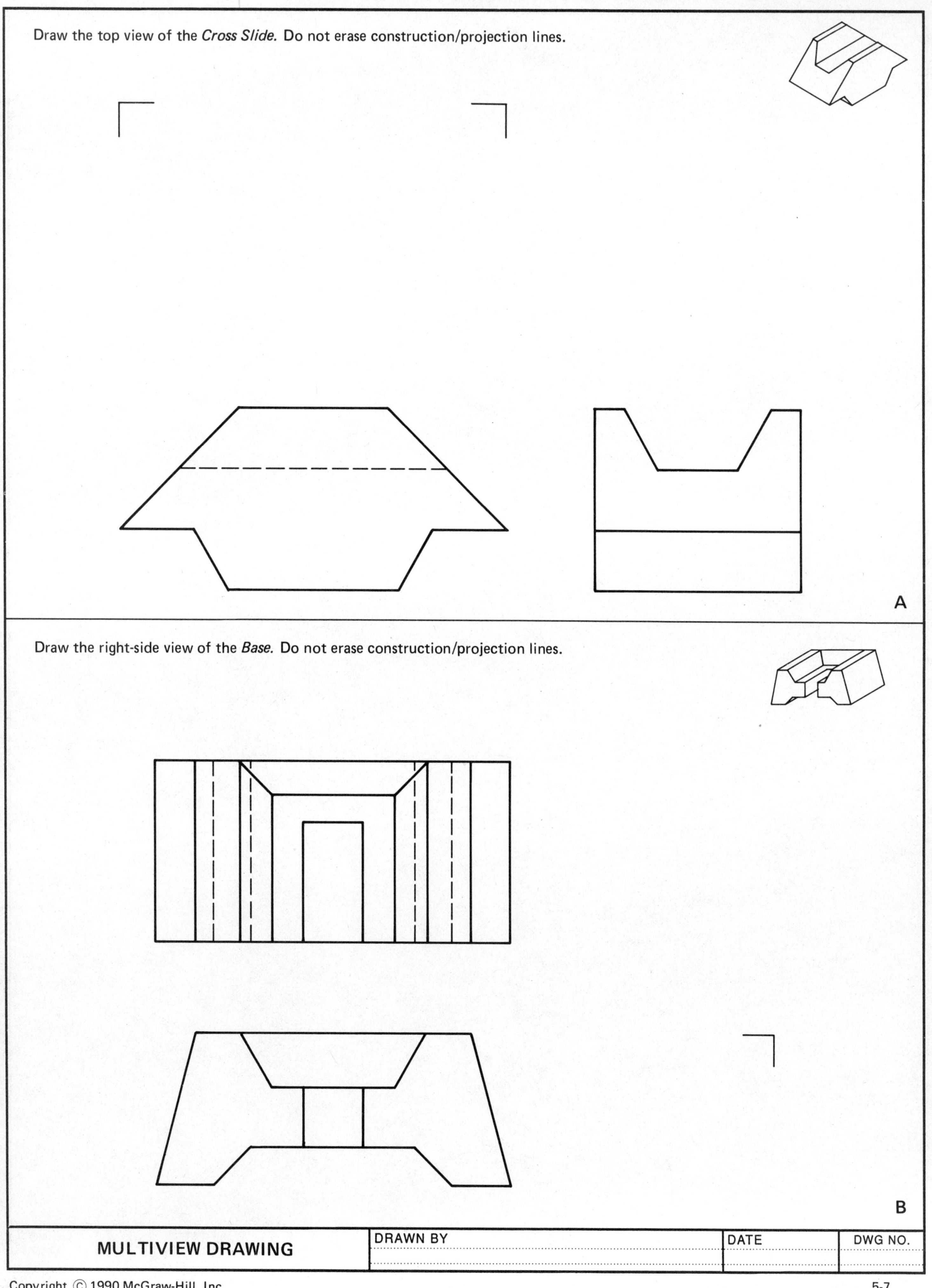
Draw the top view of the *Cross Slide.* Do not erase construction/projection lines.
A
Draw the right-side view of the *Base.* Do not erase construction/projection lines.
B
MULTIVIEW DRAWING
DRAWN BY
DATE
DWG NO.

DRAWN BY
DATE
DWG NO.

Draw the front view of the *Secondary Guide Lug.*
Do not erase construction/projection lines.

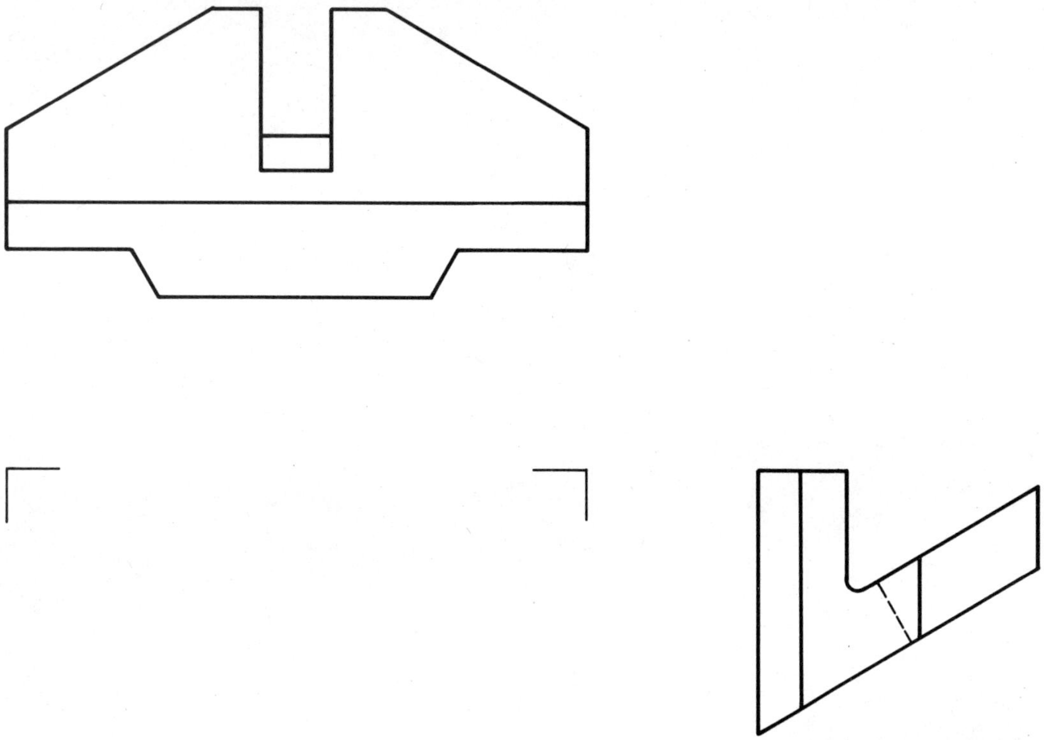

A

Draw the top view of the *Separator.*
Do not erase construction/projection lines.

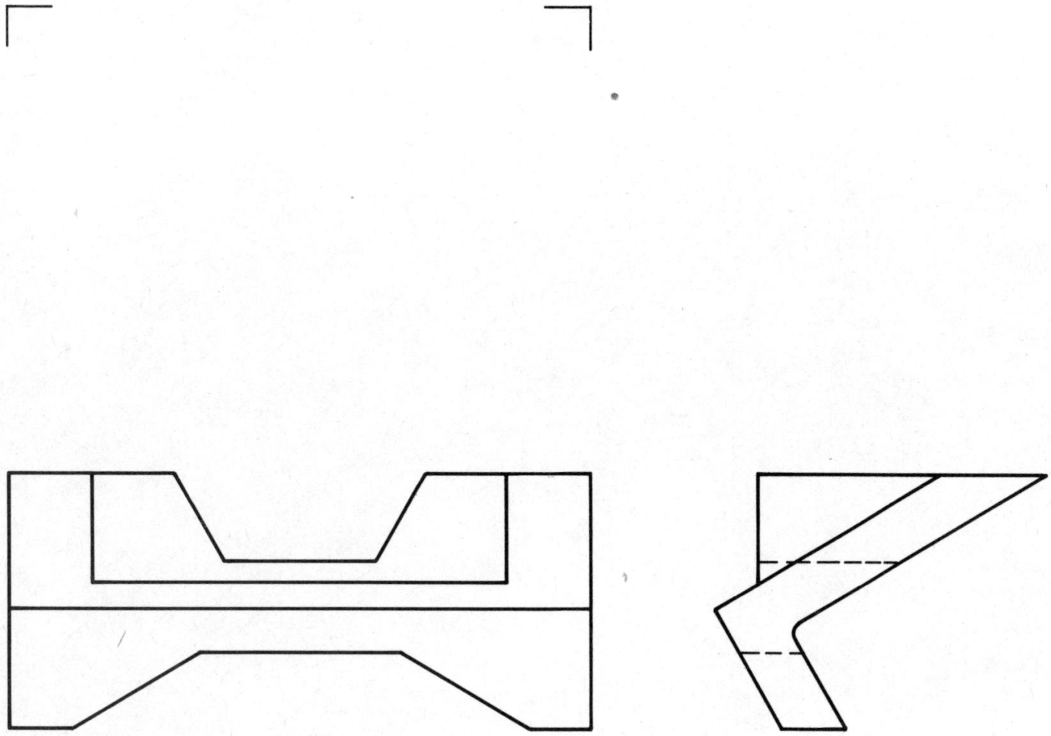

B

MULTIVIEW DRAWING	DRAWN BY	DATE	DWG NO.

	DRAWN BY	DATE	DWG NO.

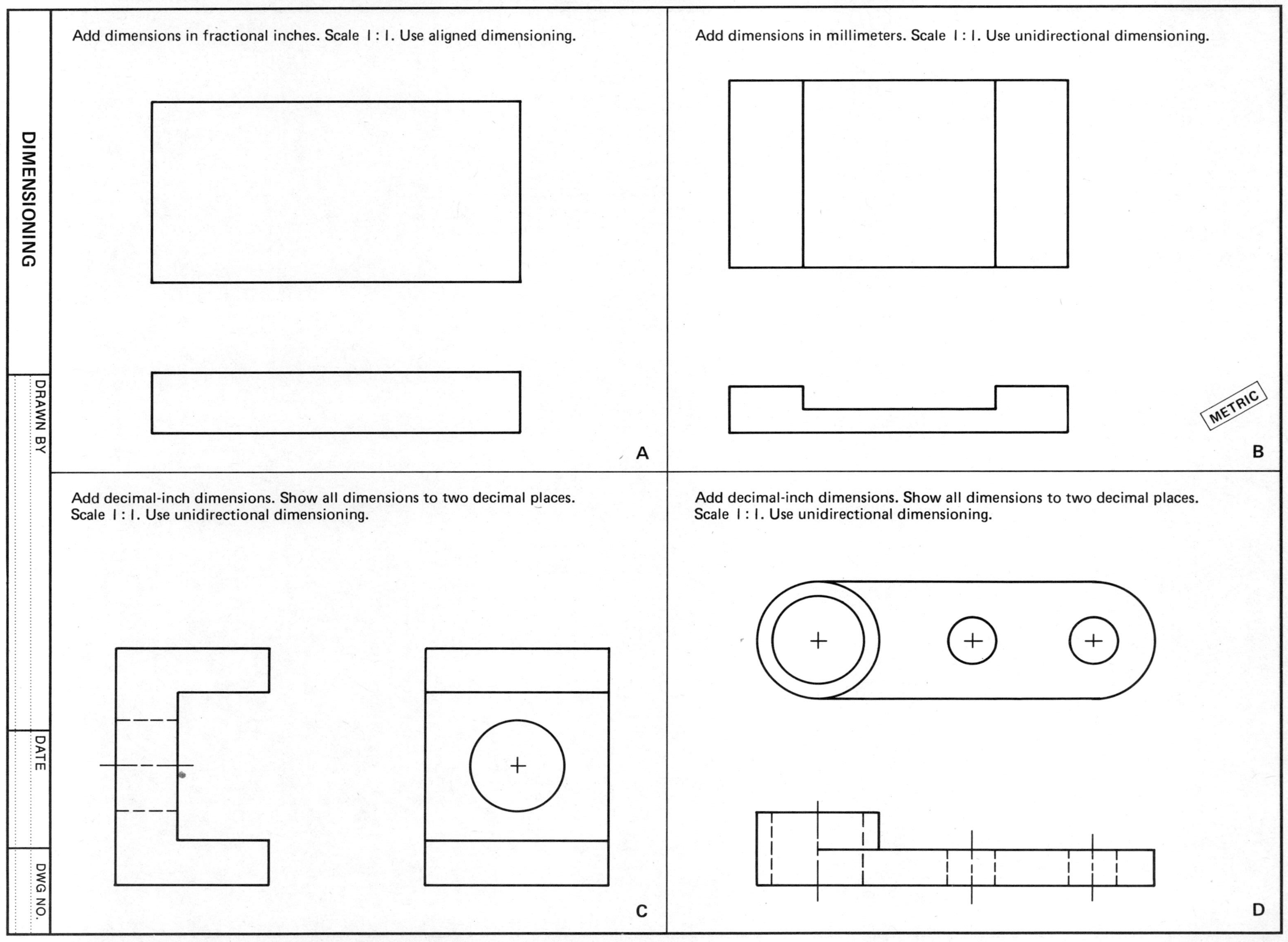
Add dimensions in fractional inches. Scale 1 : 1. Use aligned dimensioning.
A
Add dimensions in millimeters. Scale 1 : 1. Use unidirectional dimensioning.
METRIC
B
Add decimal-inch dimensions. Show all dimensions to two decimal places. Scale 1 : 1. Use unidirectional dimensioning.
C
Add decimal-inch dimensions. Show all dimensions to two decimal places. Scale 1 : 1. Use unidirectional dimensioning.
D
DIMENSIONING
DRAWN BY
DATE
DWG NO.

DRAWN BY
DATE
DWG NO.

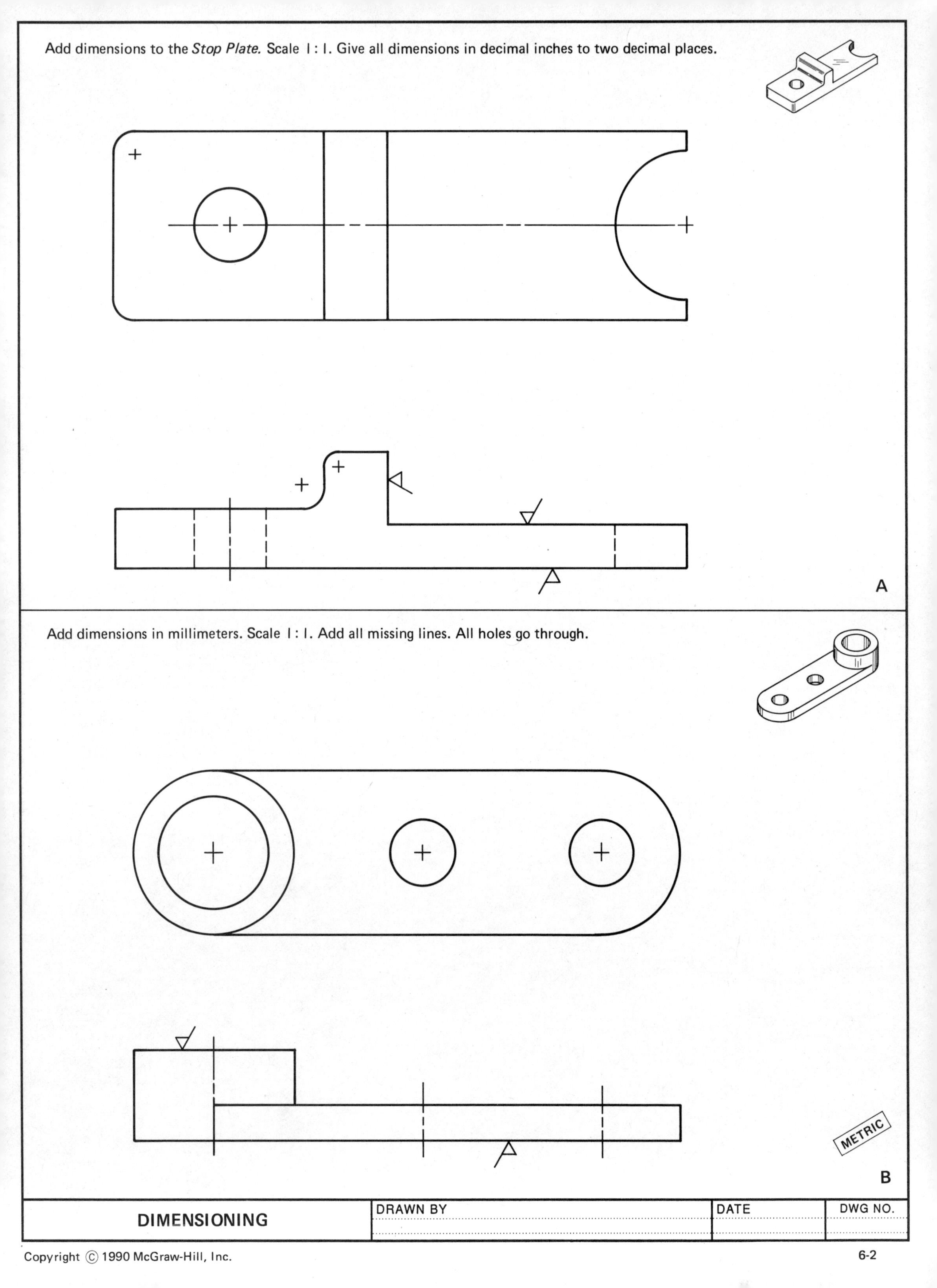
Add dimensions to the *Stop Plate.* Scale 1 : 1. Give all dimensions in decimal inches to two decimal places.
A
Add dimensions in millimeters. Scale 1 : 1. Add all missing lines. All holes go through.
METRIC
B
DIMENSIONING
DRAWN BY
DATE
DWG NO.
Copyright © 1990 McGraw-Hill, Inc.
6-2

	DRAWN BY	DATE	DWG NO.

Add dimensions in millimeters. Scale 1 : 1. Fillets and rounds R3. Add any missing lines.

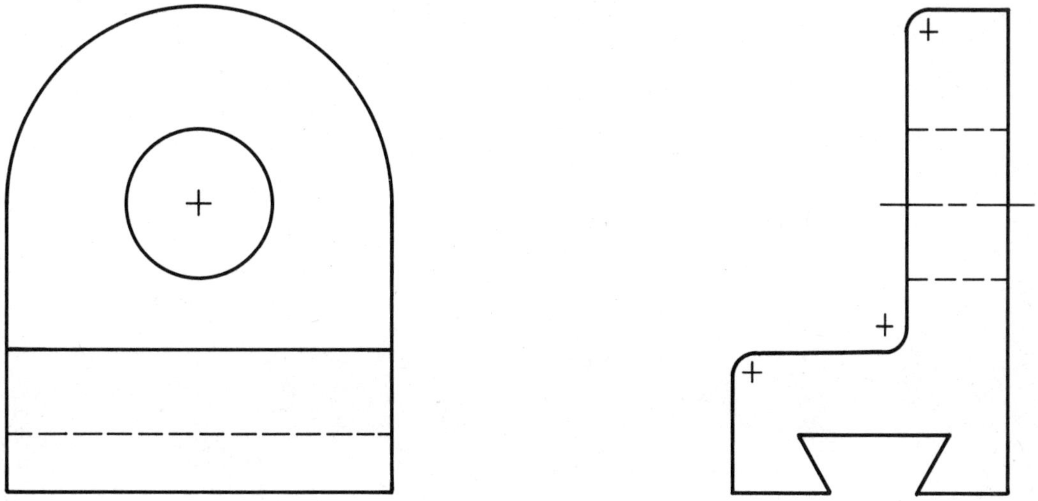

METRIC

A

Add dimensions in decimal inches. Give all dimensions in two decimal places. Scale 1 : 1. Add any missing lines.

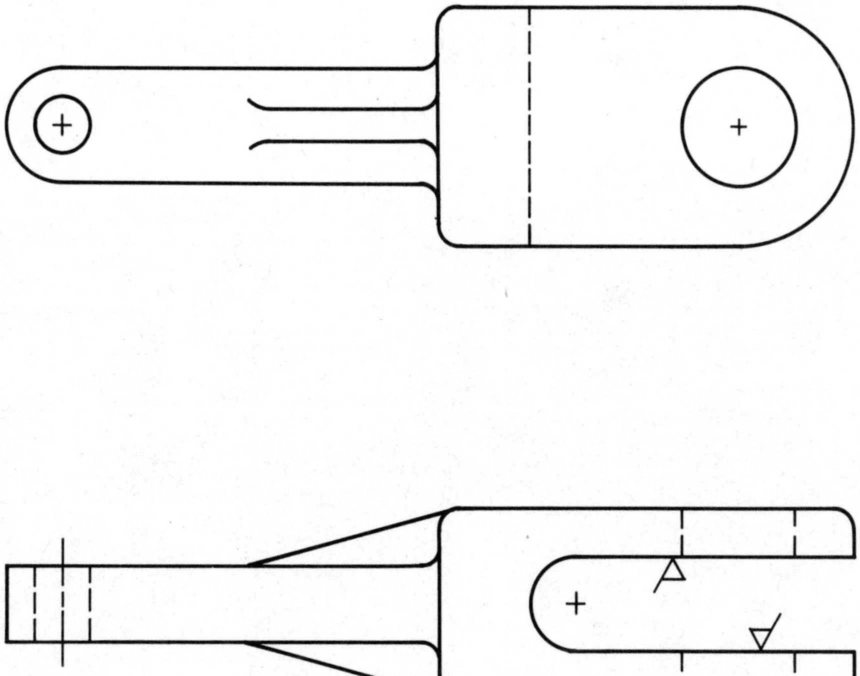

B

DIMENSIONING	DRAWN BY	DATE	DWG NO.

DRAWN BY
DATE
DWG NO.

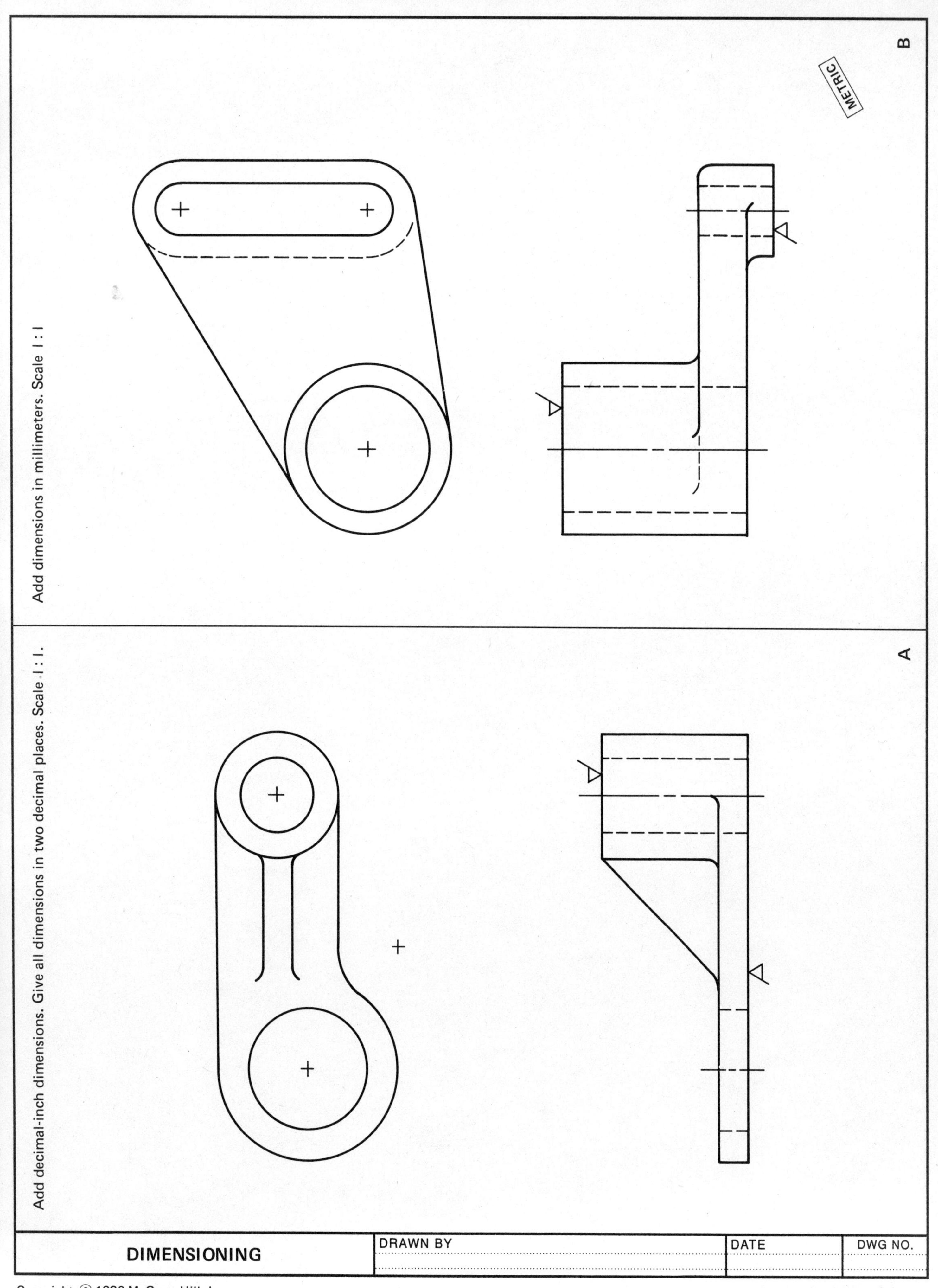
Add dimensions in millimeters. Scale 1 : 1
METRIC
B
Add decimal-inch dimensions. Give all dimensions in two decimal places. Scale 1 : 1.
A
DIMENSIONING
DRAWN BY
DATE
DWG NO.

DRAWN BY
DATE
DWG NO.

Take information from this page and use the tables on Limits and Fits in the Appendix of your textbook to complete the table on the adjoining page. All fits are calculated using the basic hole system. Item A is completed for your information.

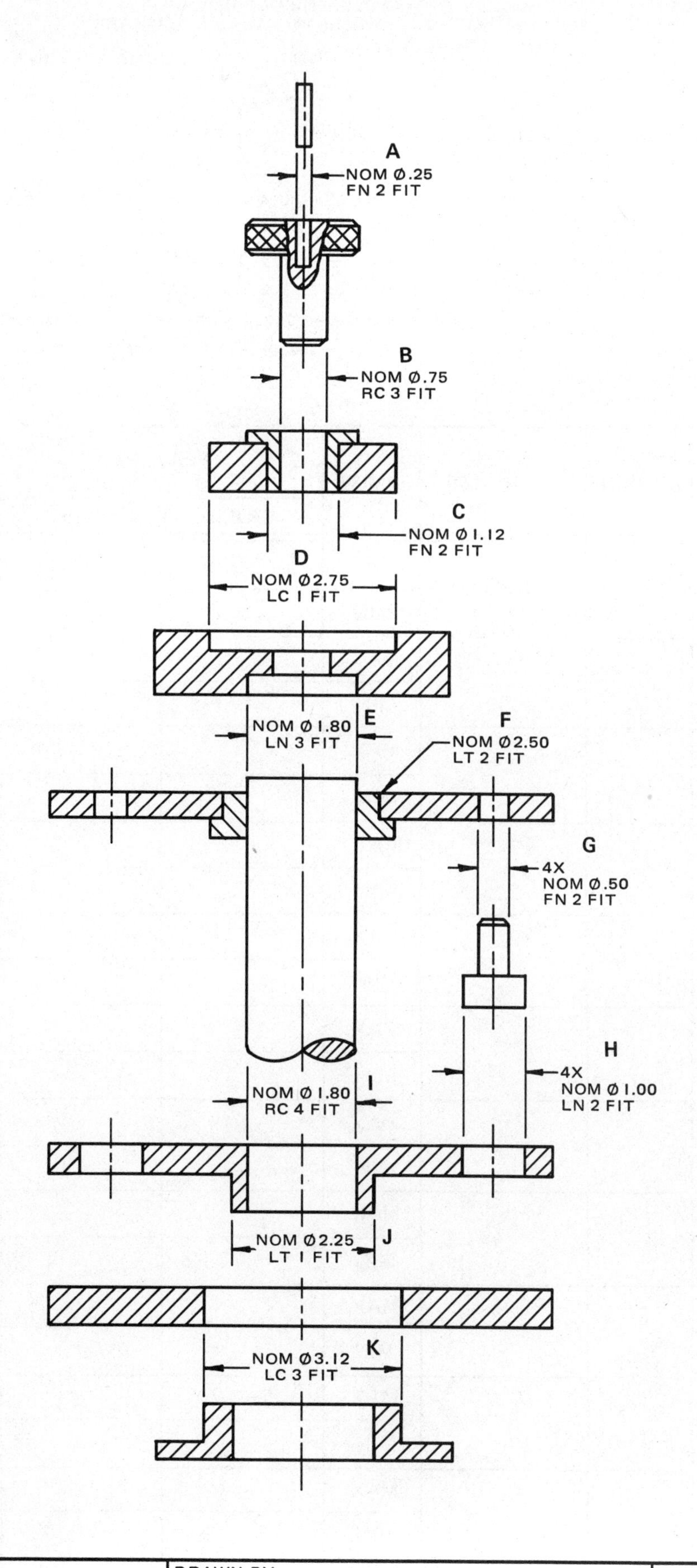

DIMENSIONING	DRAWN BY	DATE	DWG NO.

DIMENSIONING MATING PARTS

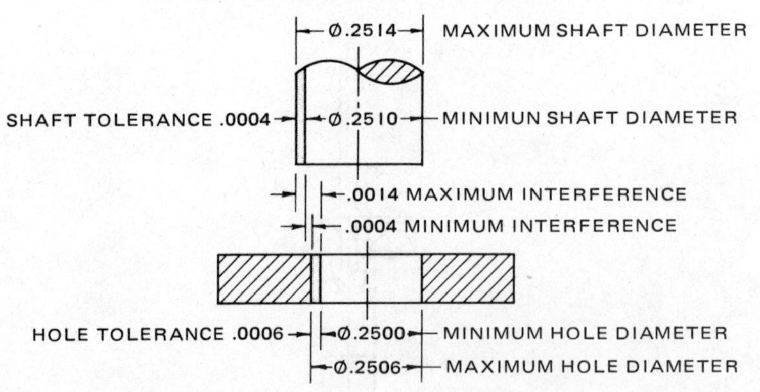

ITEM	NOMINAL SIZE	SYMBOL	BASIC SIZE	LIMITS	FEATURE		CLEARANCE OR INTERFERENCE	
					HOLE	SHAFT	MAX	MIN
A	Ø.25	FN 2	Ø.2500	MAX	.2506	.2514	.0014	.0004
				MIN	.2500	.2510		
B				MAX				
				MIN				
C				MAX				
				MIN				
D				MAX				
				MIN				
E				MAX				
				MIN				
F				MAX				
				MIN				
G				MAX				
				MIN				
H				MAX				
				MIN				
I				MAX				
				MIN				
J				MAX				
				MIN				
K				MAX				
				MIN				

DIMENSIONING	DRAWN BY	DATE	DWG NO.

GEOMETRIC TOLERANCING

The letters A, B, C, and D on the pictorial drawing represent datums A, B, C, and D respectively on the two-view drawing below. Use the following information to add datums and feature control frames.

1. Datum A is to be flat within .005
2. Datum B is to be parallel to datum A within .008
3. Datums C and D are to be parallel to datum A within .002
4. The Ø.50 hole through the cylinder is to be perpendicular to datum A within Ø.005

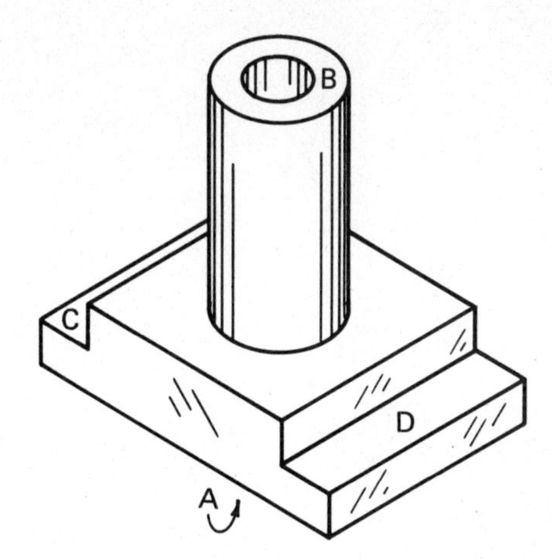

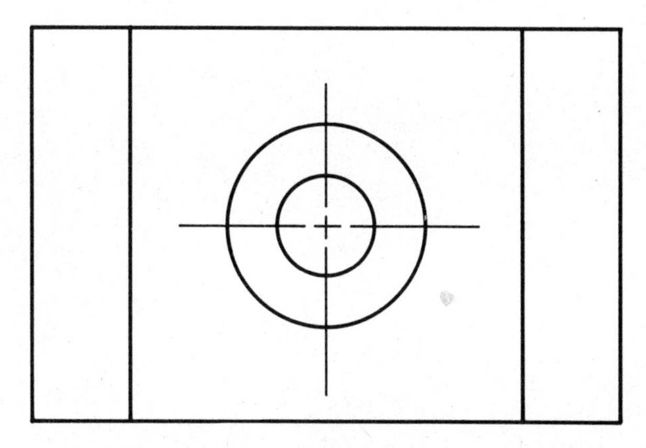

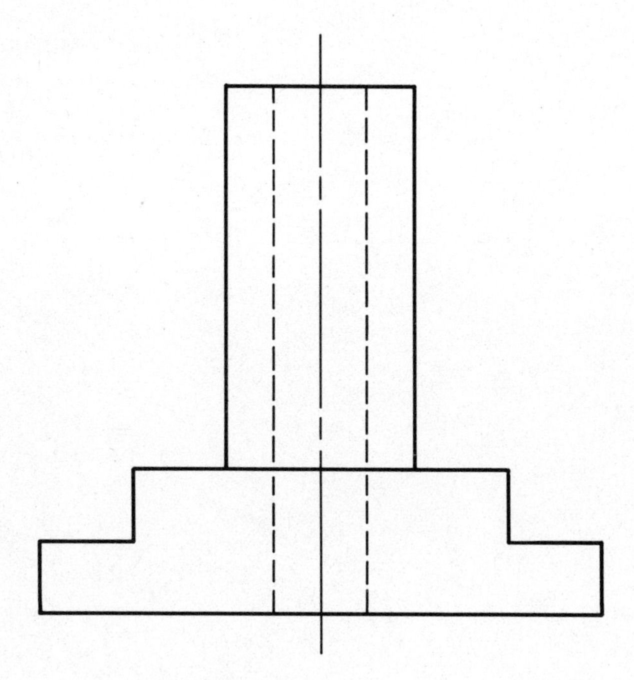

DIMENSIONING	DRAWN BY	DATE	DWG NO.

DRAWN BY
DATE
DWG NO.

Sketch the auxiliary view of surface A in each problem. Use a soft-lead pencil and try to match the line weights given.

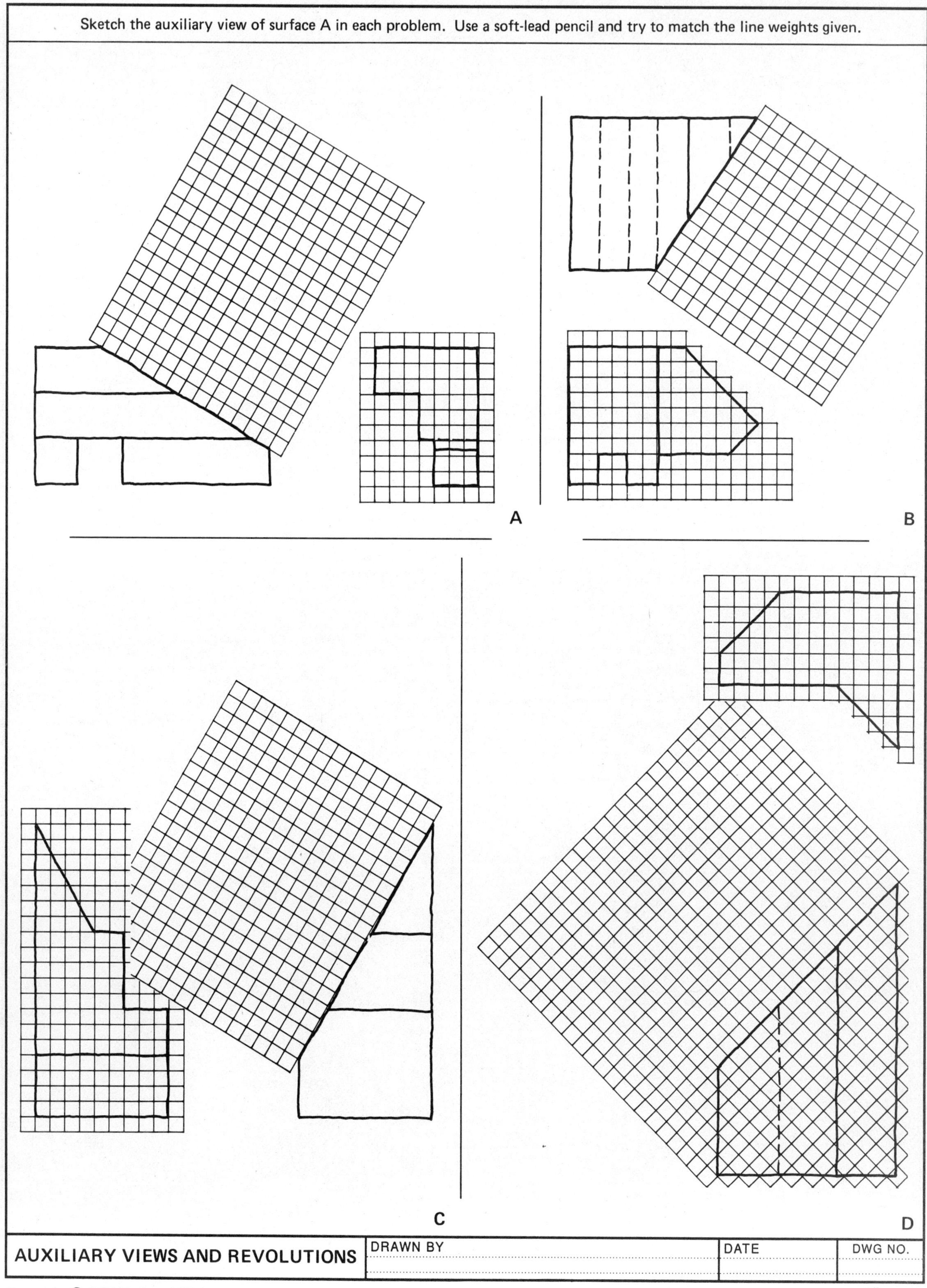

AUXILIARY VIEWS AND REVOLUTIONS

DRAWN BY	DATE	DWG NO.

DRAWN BY
DATE
DWG NO.

Create the auxiliary view of surface A in each of the problems below. Do not erase projection lines.

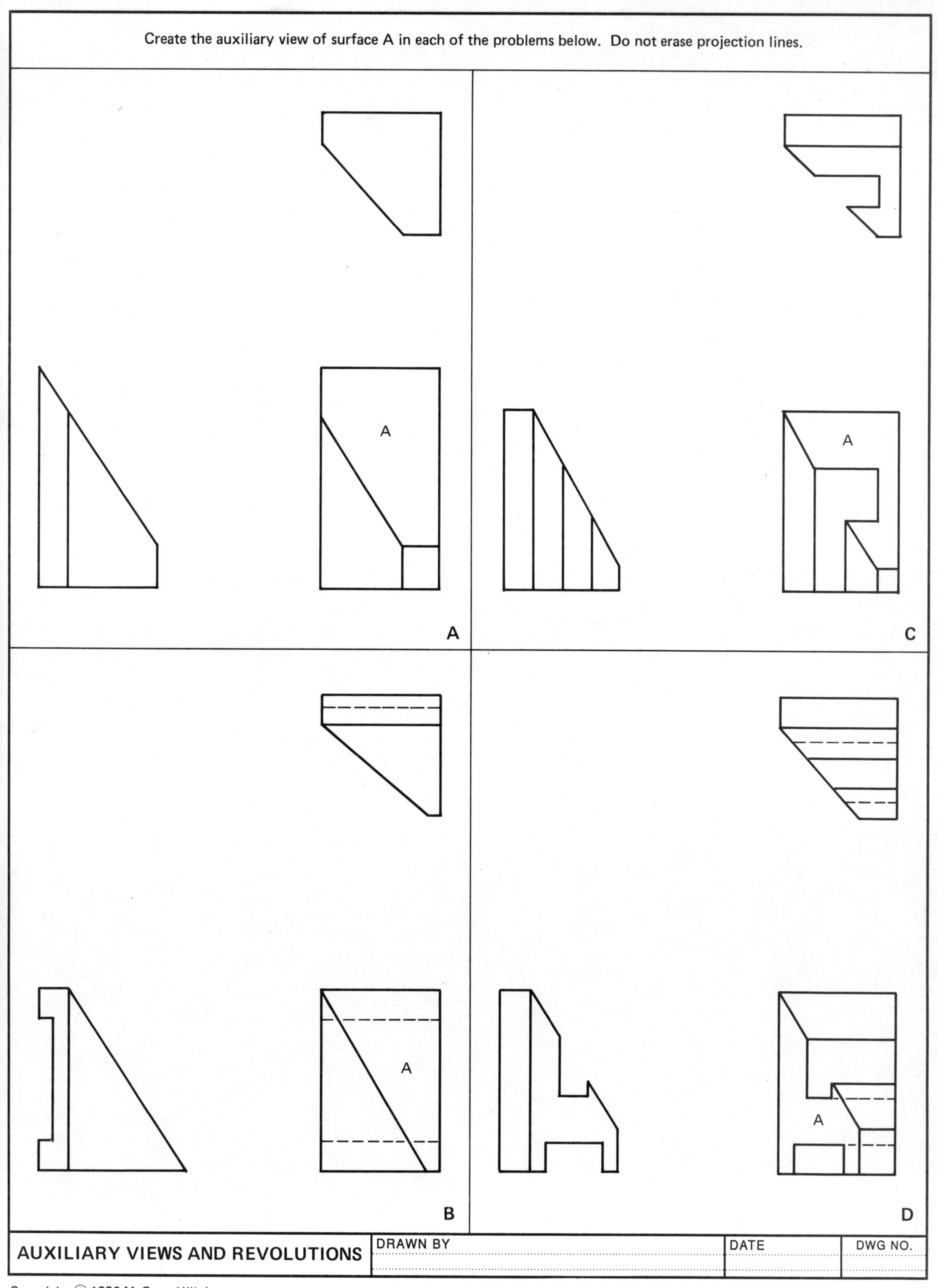

AUXILIARY VIEWS AND REVOLUTIONS

DRAWN BY	DATE	DWG NO.

	DRAWN BY	DATE	DWG NO.

Make a full-size drawing of the *Anchor Lug* including a complete auxiliary view. Convert fractions to decimals and use only current ANSI standards for dimensioning. Do not erase construction lines.

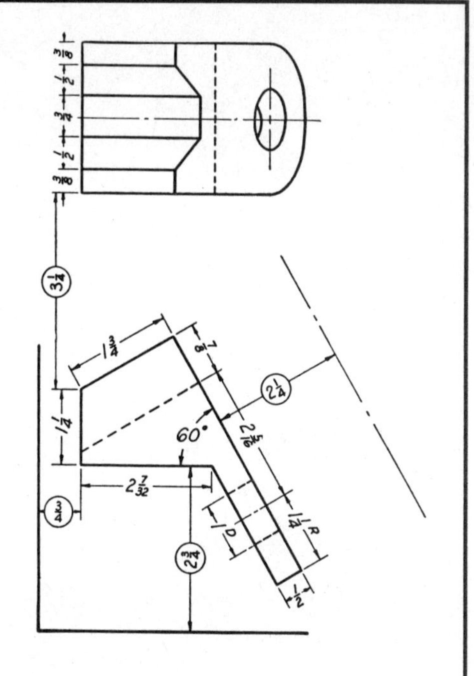

AUXILIARY VIEWS AND REVOLUTIONS	DRAWN BY	DATE	DWG NO.

	DRAWN BY	DATE	DWG NO.

Make a drawing of the *Inclined Bearing.* Include an auxiliary view of the inclined surface. Scale 1 : 1. Dimensions optional.

INCLINED STOP

AUXILIARY VIEWS AND REVOLUTIONS

DRAWN BY

DATE

DWG NO.

DRAWN BY
DATE
DWG NO.

The front and partial right-side views of an *Inclined Stop* are shown. Draw an auxiliary view to show the true shape of the inclined surface and dimension the auxiliary view. Scale 1 : 1. Match line weights as closely as possible.

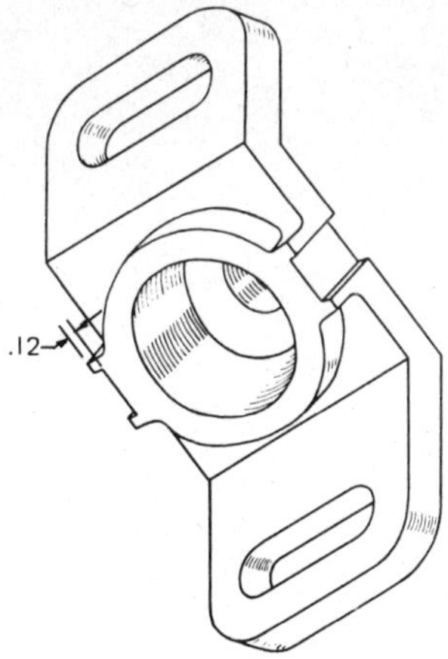

.12

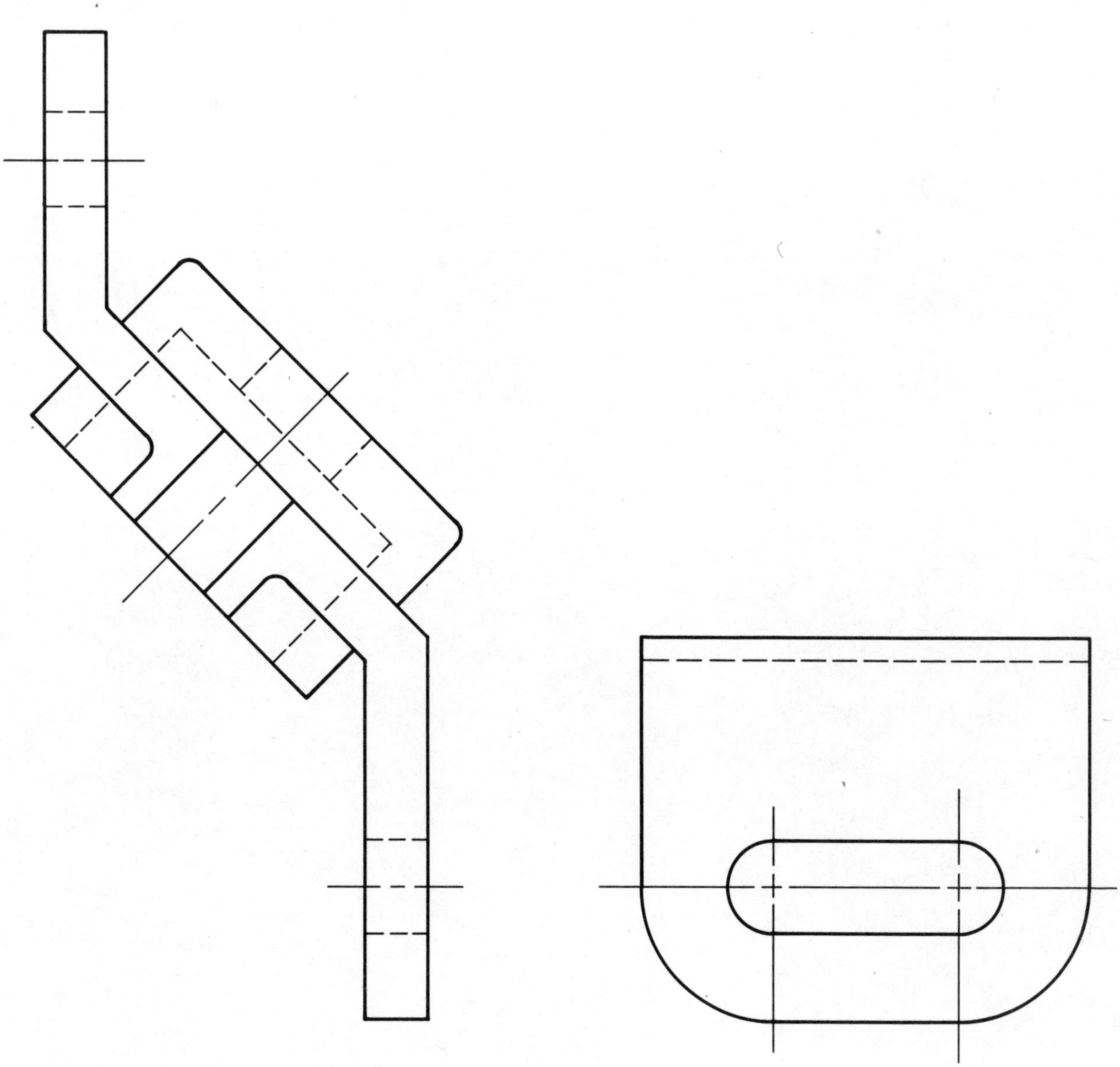

AUXILIARY VIEWS AND REVOLUTIONS	DRAWN BY	DATE	DWG NO.

DRAWN BY
DATE
DWG NO.

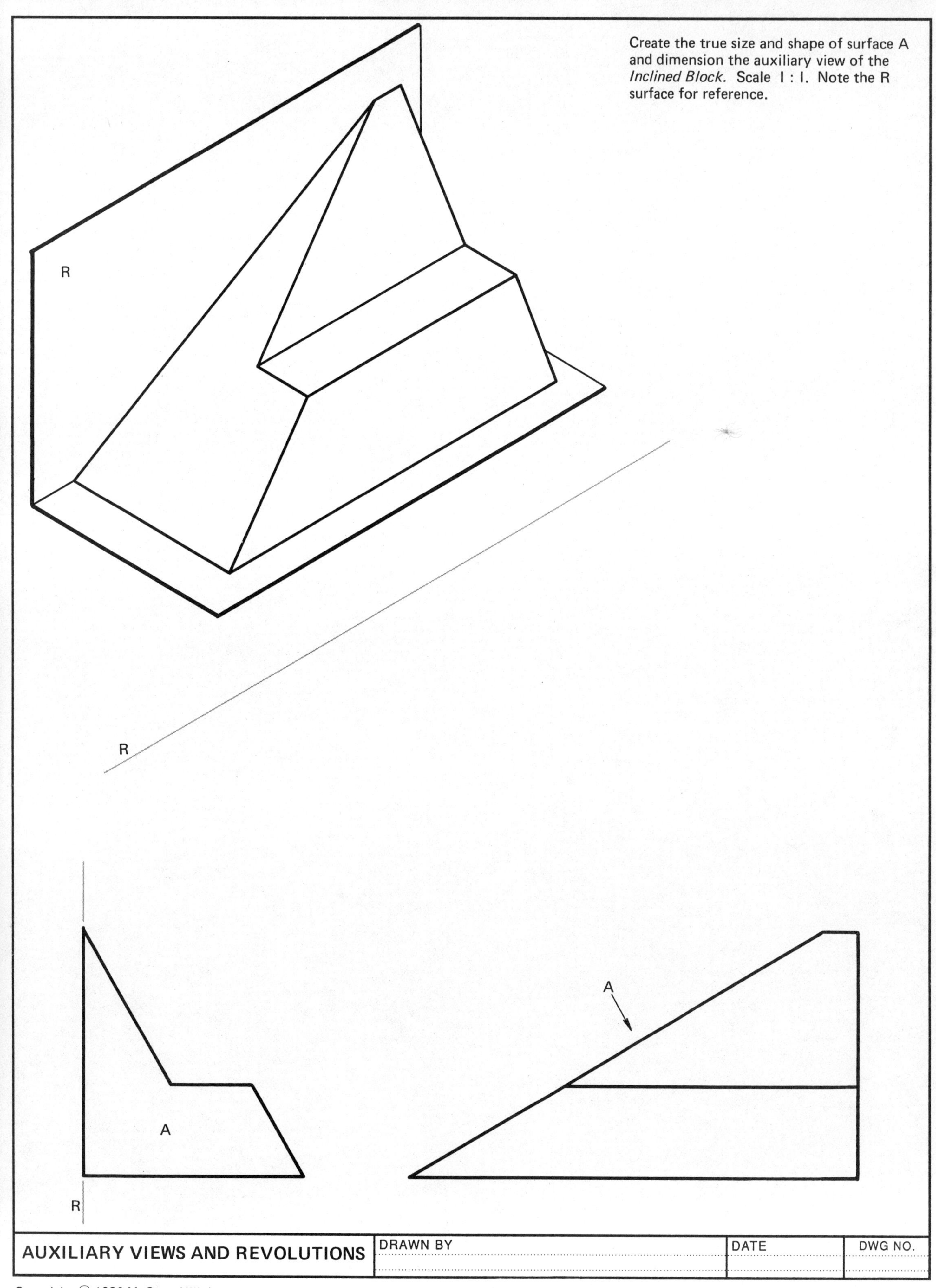

AUXILIARY VIEWS AND REVOLUTIONS

DRAWN BY	DATE	DWG NO.

DRAWN BY
DATE
DWG NO.

Given the left-side view and top view, develop the top-auxiliary view of the *Concave Block*. Scale 1 : 1. Note the reference plane.

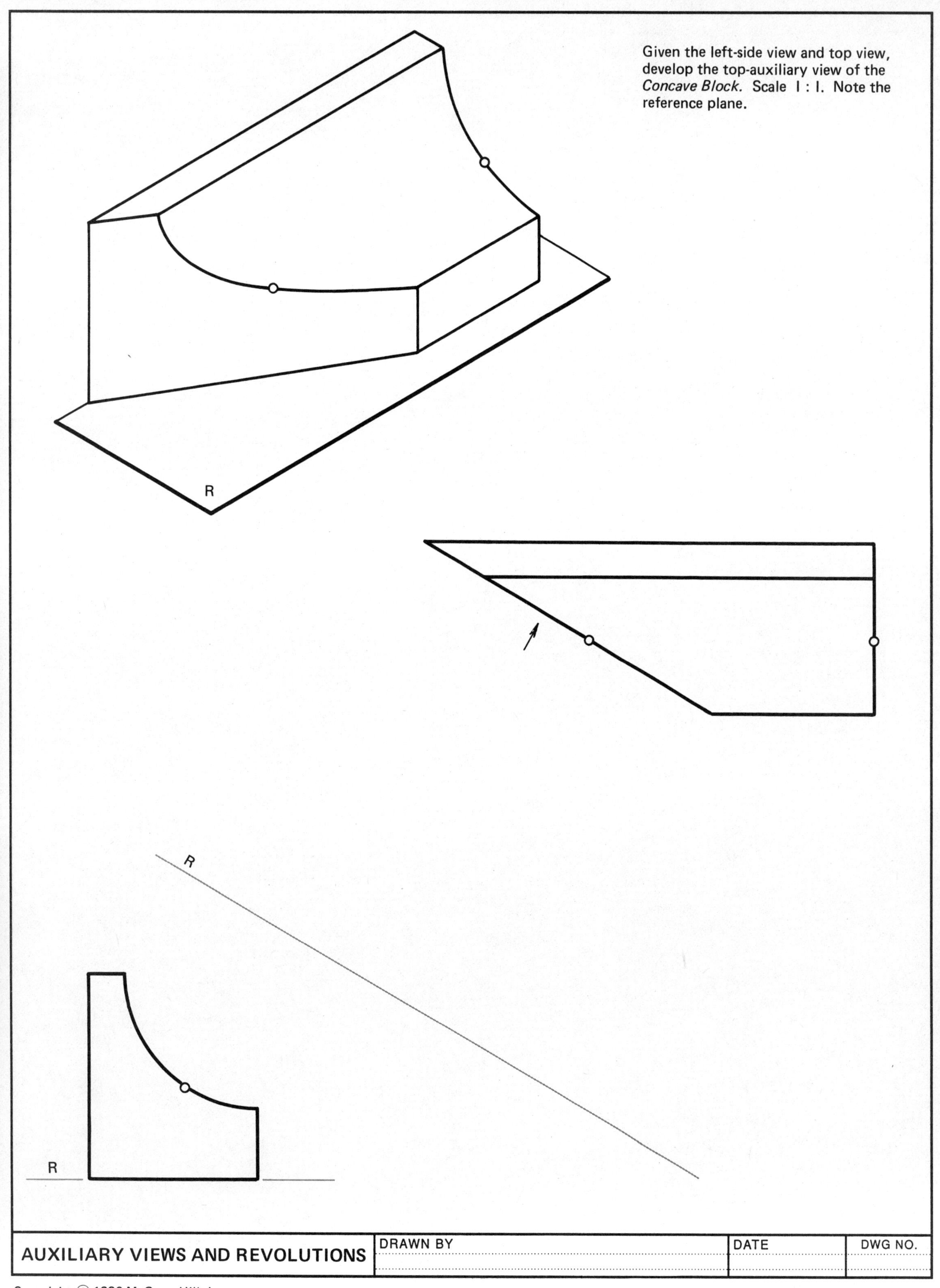

AUXILIARY VIEWS AND REVOLUTIONS	DRAWN BY	DATE	DWG NO.

DRAWN BY
DATE
DWG NO.

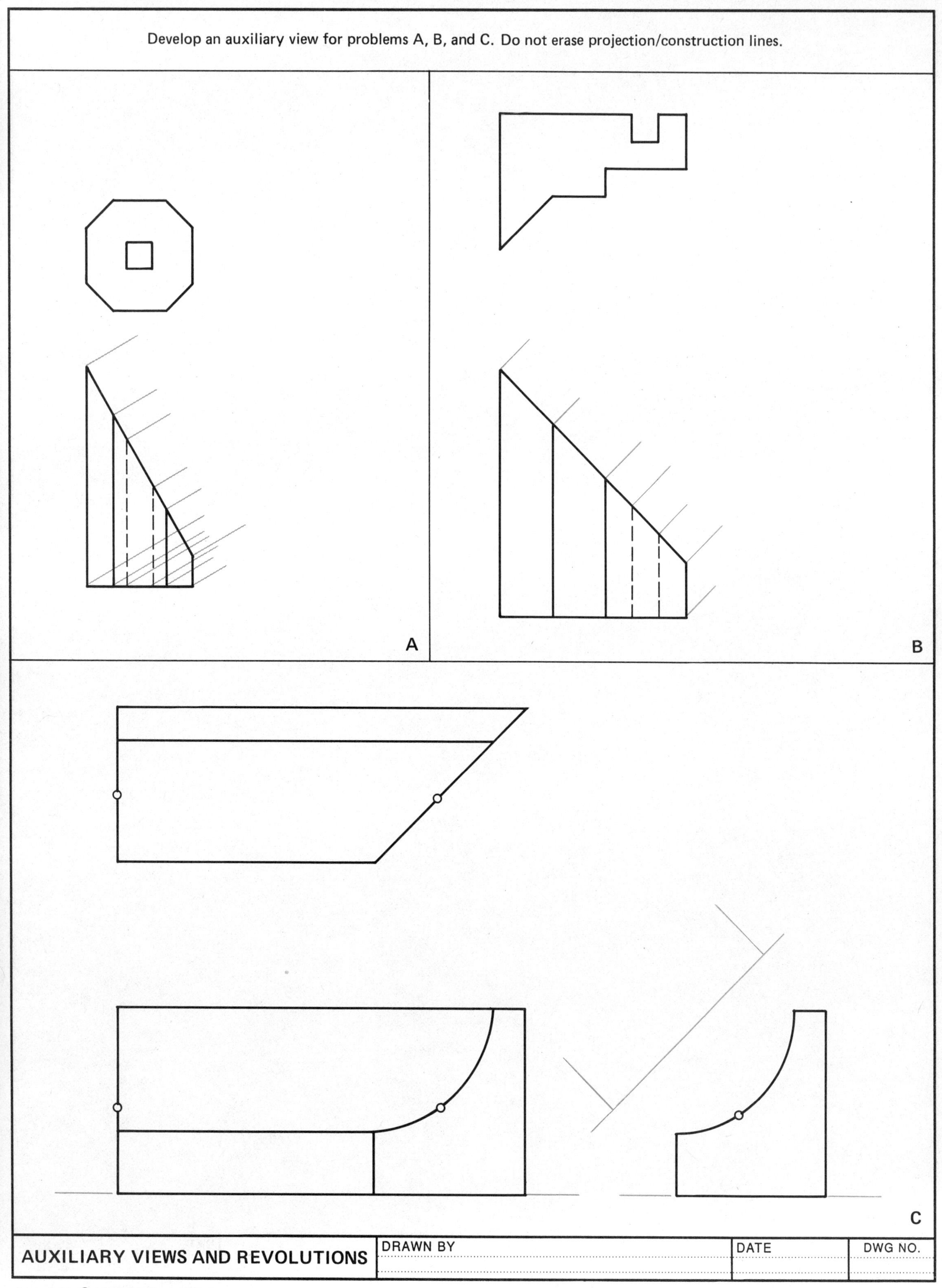

Develop an auxiliary view for problems A, B, and C. Do not erase projection/construction lines.
A
B
C
AUXILIARY VIEWS AND REVOLUTIONS
DRAWN BY
DATE
DWG NO.

DRAWN BY
DATE
DWG NO.

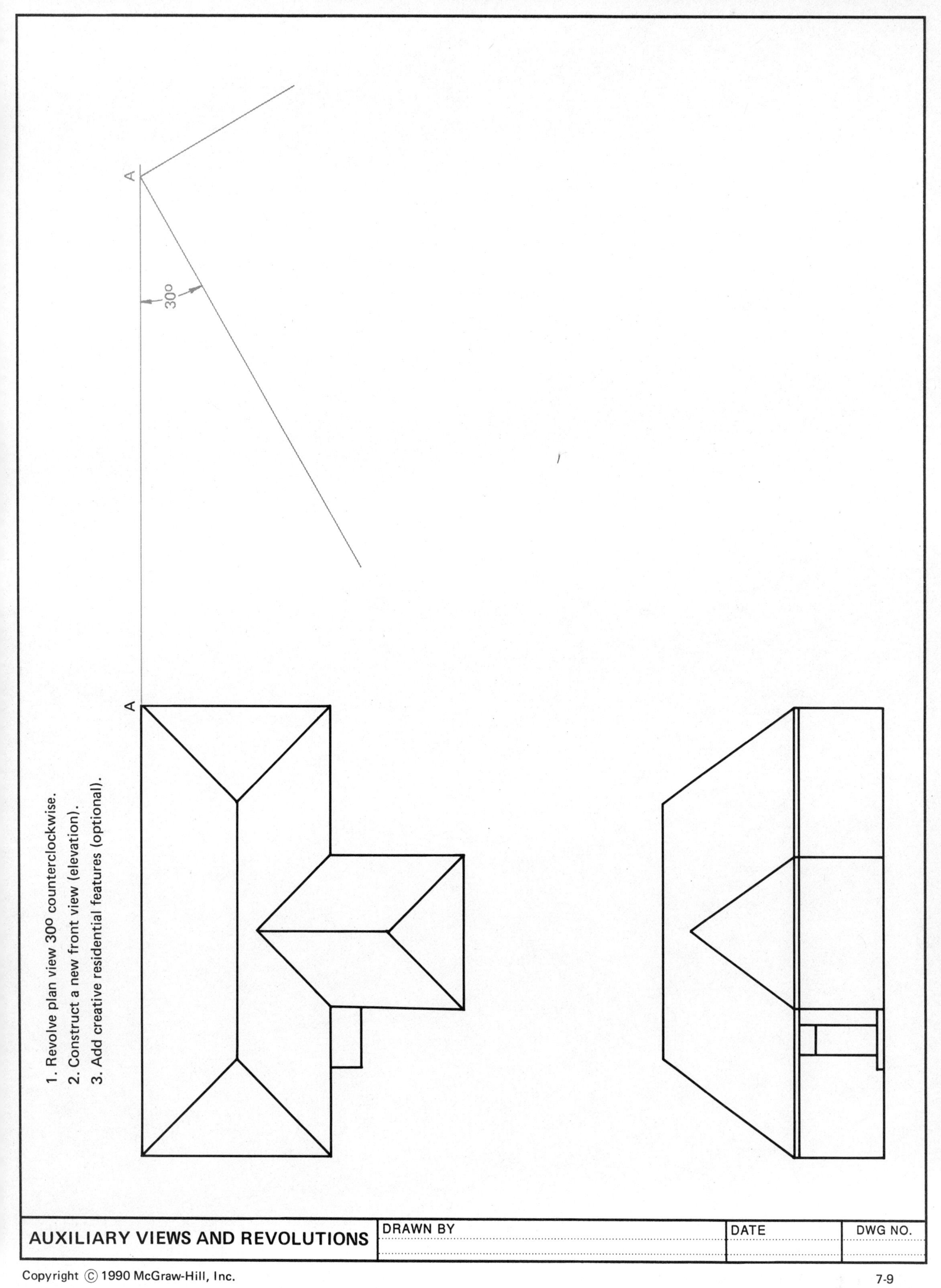
1. Revolve plan view 30° counterclockwise.
2. Construct a new front view (elevation).
3. Add creative residential features (optional).
A
A
30°
AUXILIARY VIEWS AND REVOLUTIONS
DRAWN BY
DATE
DWG NO.

	DRAWN BY	DATE	DWG NO.

Using the three normal views shown below, complete the assignments given in blocks B, C, and D. Review the unit on revolutions in your textbook before beginning this assignment. Do not erase projection lines.

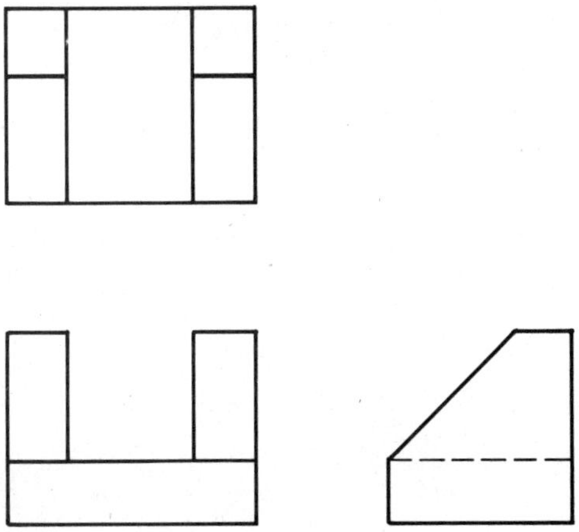

A

With the front view revolved 30°, develop the top and right-side views.

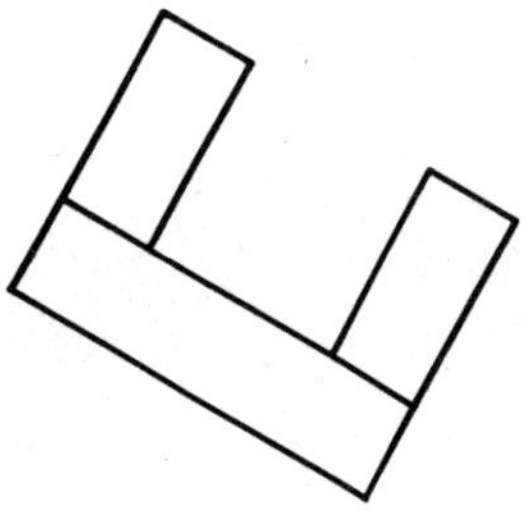

B

With the top view revolved 45°, develop the front and right-side views.

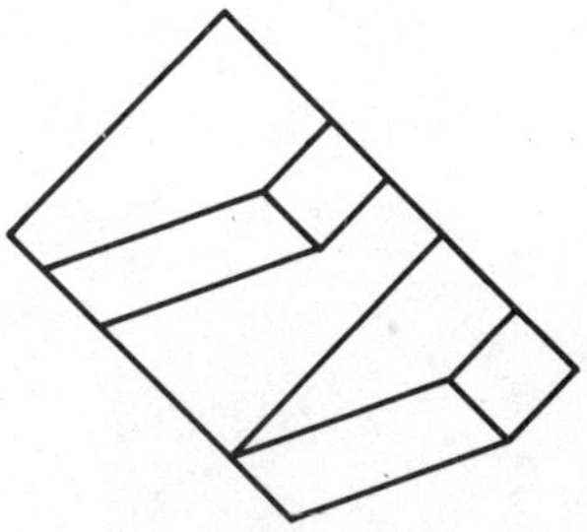

C

With the right-side view revolved 45°, develop the front and top views.

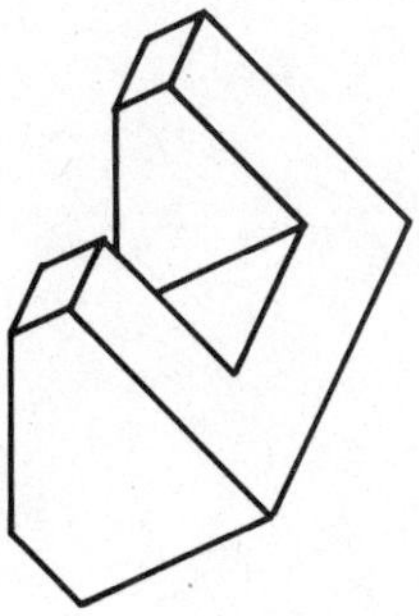

D

AUXILIARY VIEWS AND REVOLUTIONS	DRAWN BY	DATE	DWG NO.

	DRAWN BY	DATE	DWG NO.

Draw a partial auxiliary view to show the true shape of the sloping surface of the *Wedge Block.*
Scale 1 : 1. Dimensions optional. Do not erase projection lines.

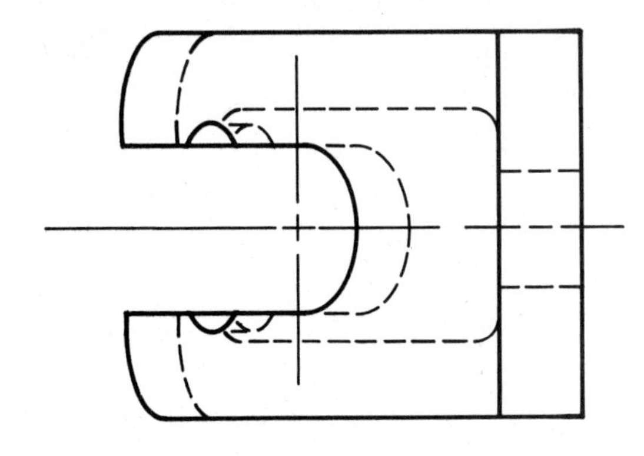

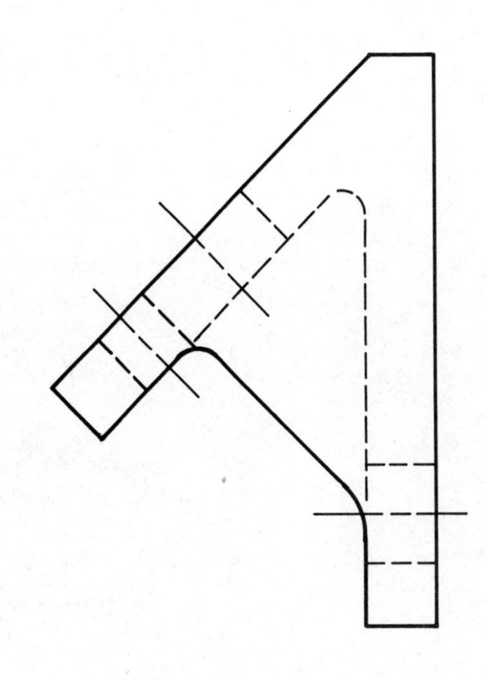

AUXILIARY VIEWS AND REVOLUTIONS	DRAWN BY	DATE	DWG NO.

	DRAWN BY	DATE	DWG NO.

Draw an auxiliary view of the *Angle Plate* to show the true shape of the inclined surface. Consult the pictorial sketch for the size and location of holes and other necessary dimensions. Complete the front view to show missing hidden lines. Estimate any sizes not given. Scale 3/4 size. Dimensions optional.

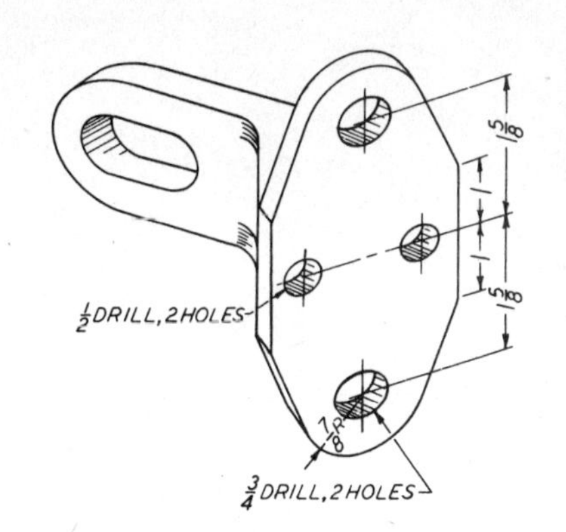

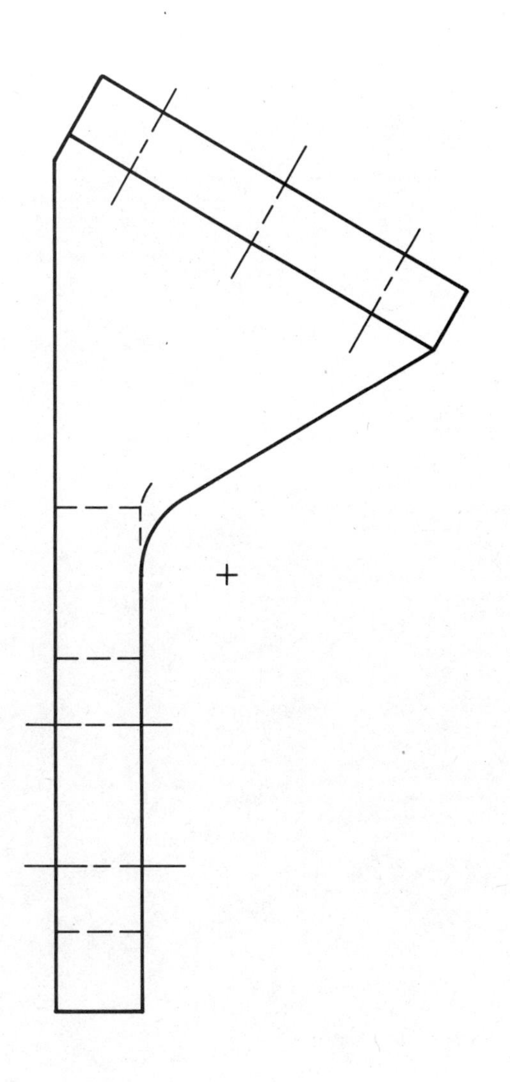

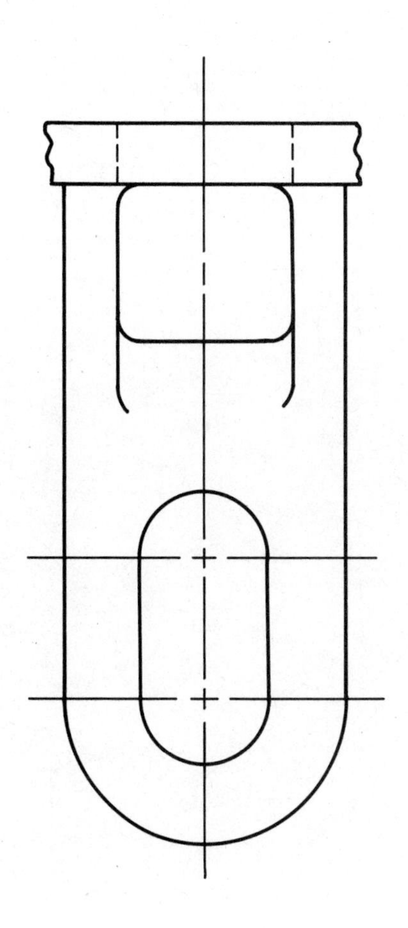

AUXILIARY VIEWS AND REVOLUTIONS	DRAWN BY	DATE	DWG NO.

DRAWN BY
DATE
DWG NO.

Create the auxiliary view and dimension the true-size projection. Develop view B—B to clarify hidden details in the top view. Scale 1 : 1.

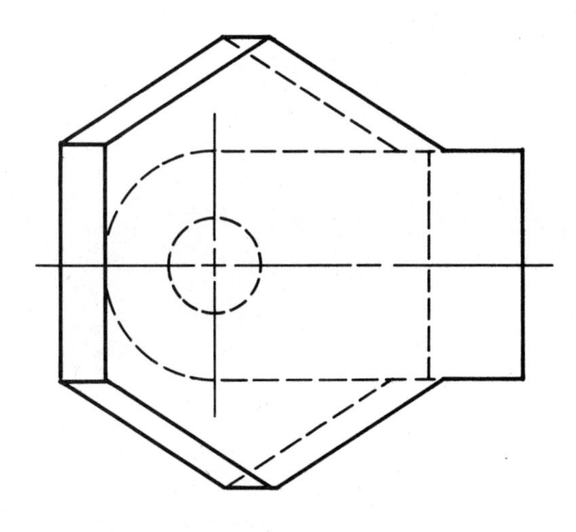

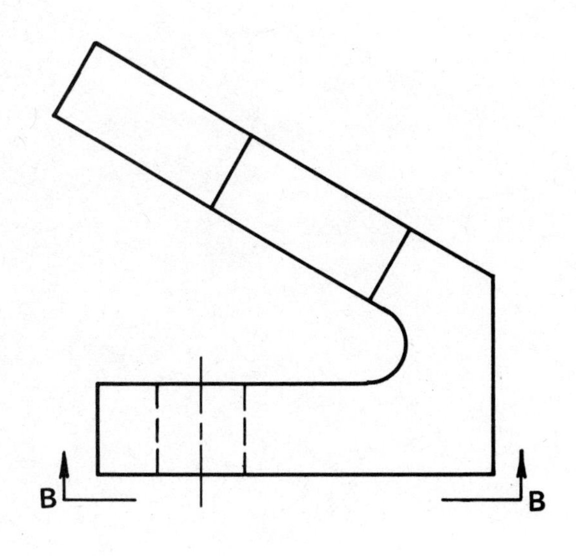

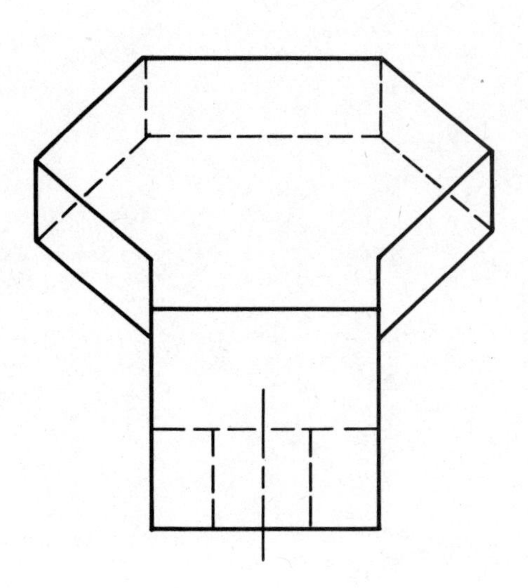

AUXILIARY VIEWS AND REVOLUTIONS	DRAWN BY	DATE	DWG NO.

	DRAWN BY	DATE	DWG NO.

Create the first and second auxiliary views of the object to develop the true size and shape of surface (plane) 0, 1, 2.

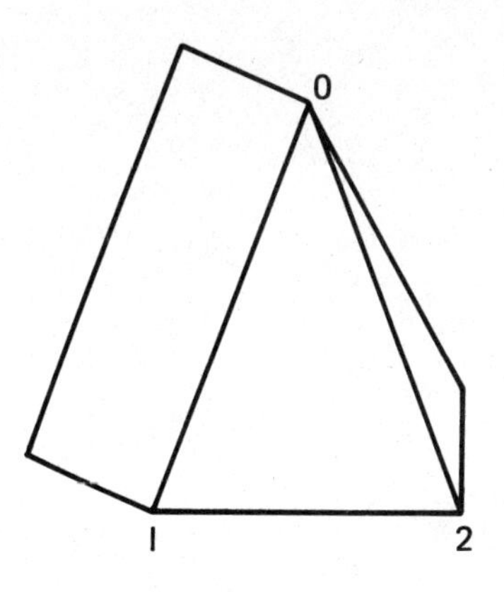

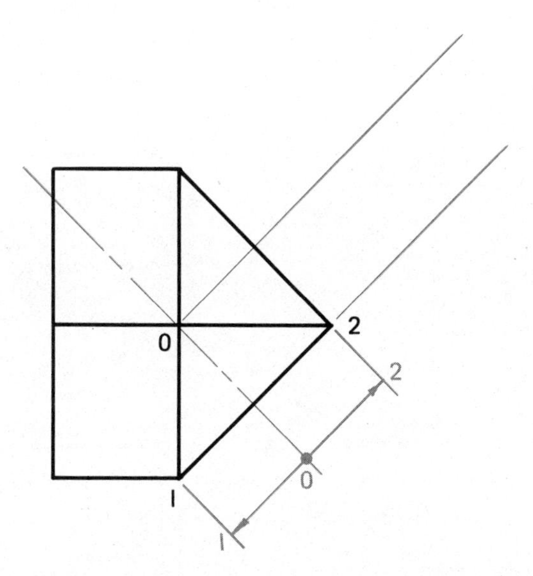

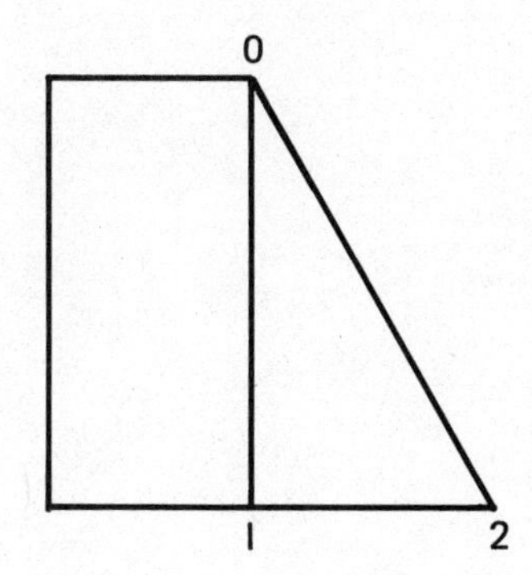

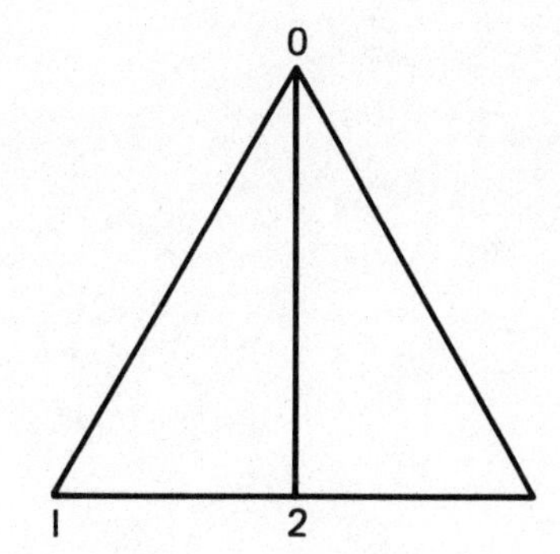

AUXILIARY VIEWS AND REVOLUTIONS

DRAWN BY	DATE	DWG NO.

DRAWN BY
DATE
DWG NO.

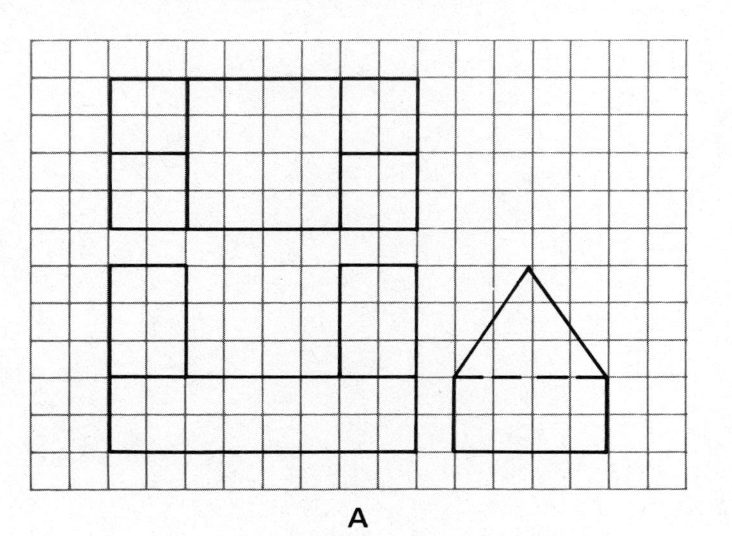

A

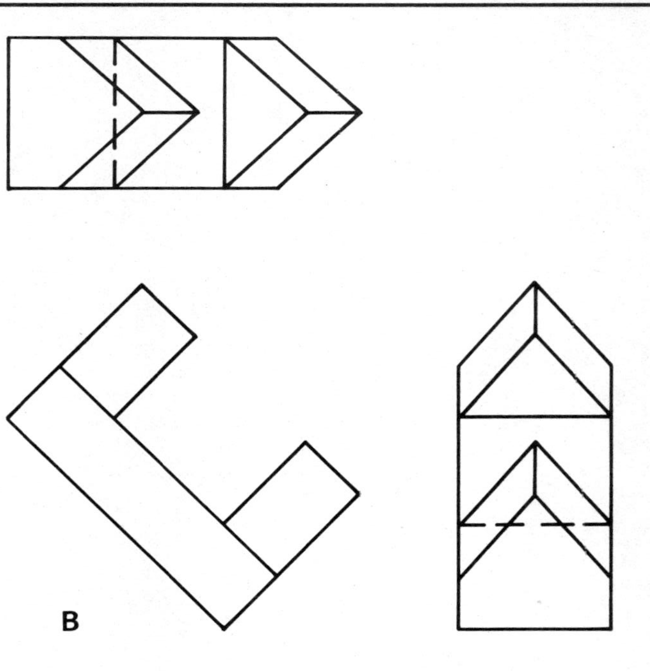

B

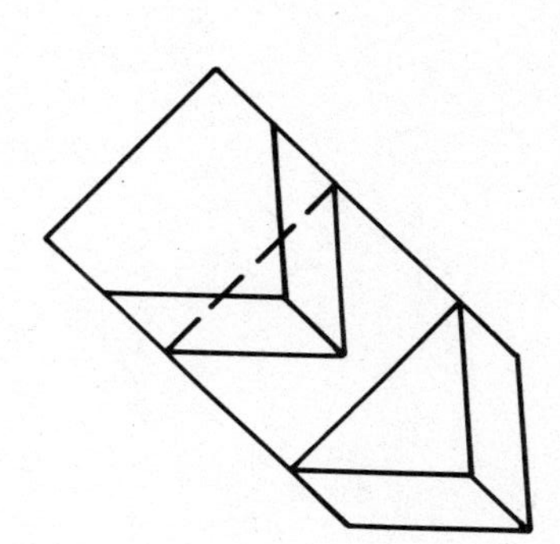

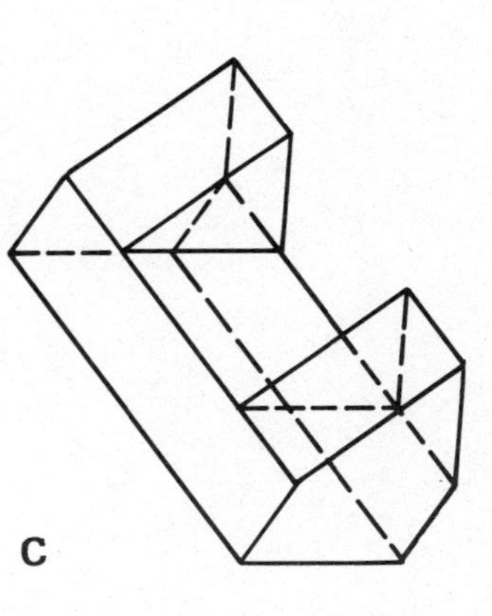

C

Three normal views of the object are shown at A. Figure B shows the object with the front view revolved 45° and the top and right-side views projected from it. At C, the top view is rotated 45° clockwise from its position in Figure B and the front view is projected from it.

Develop the right-side view by projecting points from the top and front views. Additional successive revolutions can be accomplished simply by rotating any of the views.

Study Figures 7-23 through 7-26 in your textbook before beginning this assignment.

AUXILIARY VIEWS AND REVOLUTIONS	DRAWN BY	DATE	DWG NO.

DRAWN BY
DATE
DWG NO.

A_T

T

F

A_F

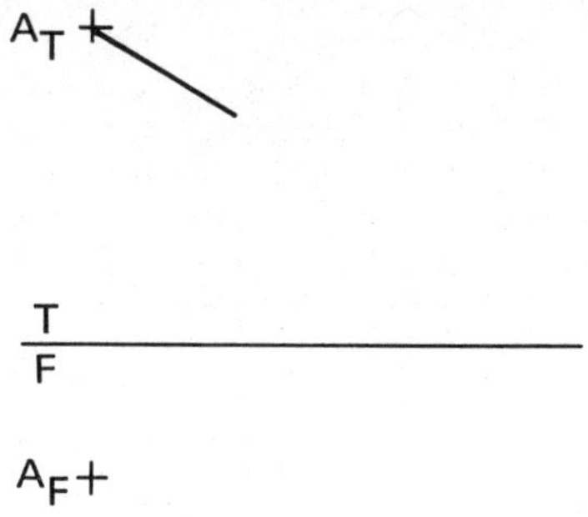

Line AB is parallel to the T plane and is 1.50 in. long. Complete the two views of the line.

Title: T.L. of Line Segment

A

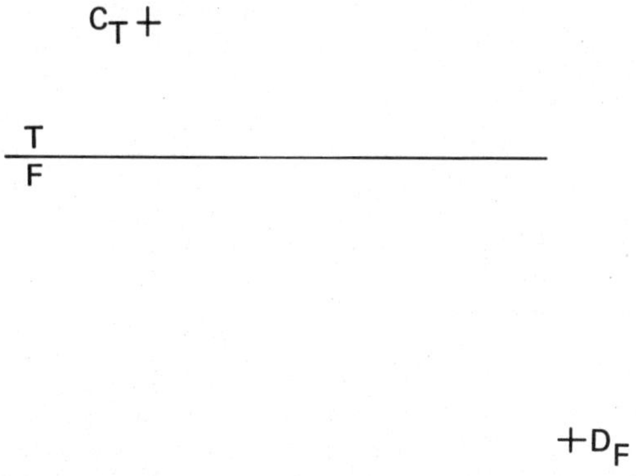

Line CD is parallel to the F plane and is 2.00 in. long. Complete the two views of the line.

C_T

T

F

D_F

Title: T.L. of Line Segment

B

T.L of AB: ______ in.

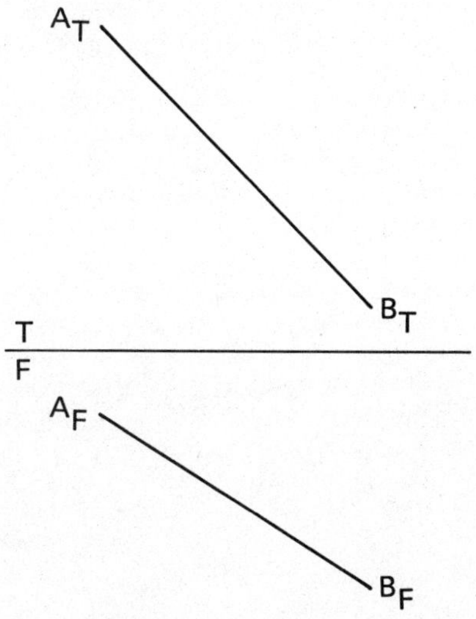

A_T

B_T

T

F

A_F

B_F

Determine the T.L. of line segment AB on an auxiliary plane. Measure the line and enter the length above. Label the auxiliary plane with a 1.

Title: T.L. of Line Segment

C

T.L. of CD: ______ in.

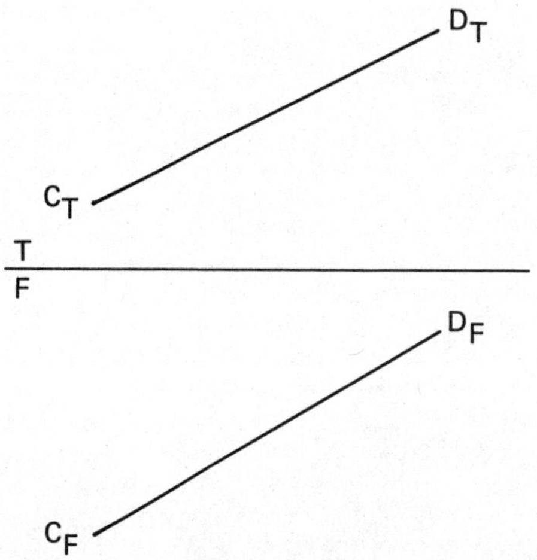

D_T

C_T

T

F

D_F

C_F

Determine the T.L of line segment CD on an auxiliary plane. Measure the line and enter the length above.

Title: T.L. of Line Segment

D

DESCRIPTIVE GEOMETRY	DRAWN BY	DATE	DWG NO.

DRAWN BY	DATE	DWG NO.

The point view of a line is shown on a plane which is perpendicular to the line.

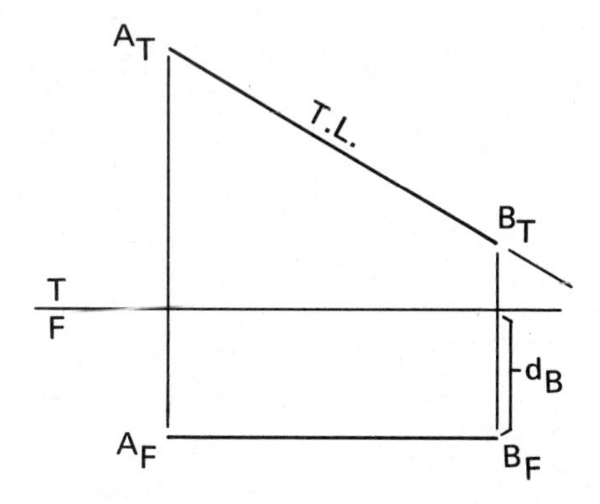

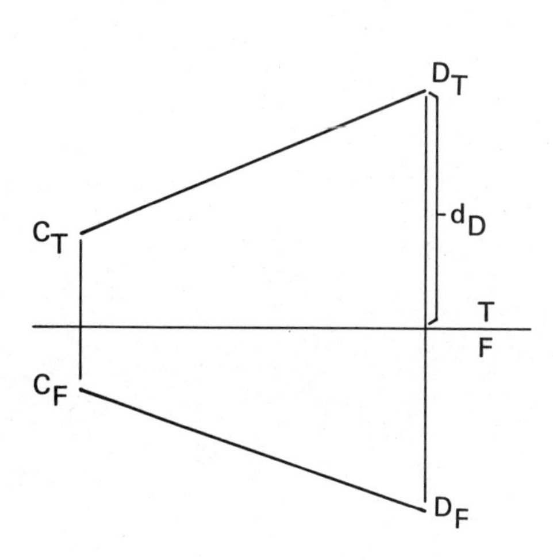

Determine the point projection of lines AB and CD.

Title: Point View of a Line

A

Determine the point view of line RS.

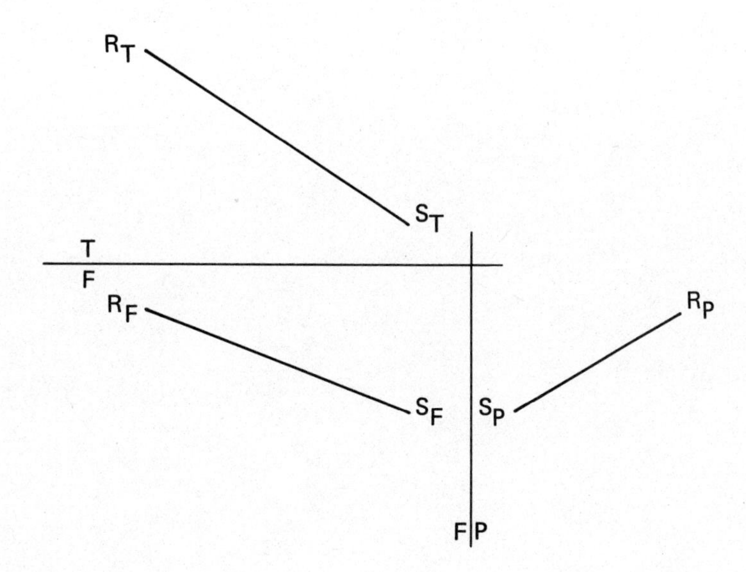

Title: Point View of a Line

B

DESCRIPTIVE GEOMETRY	DRAWN BY	DATE	DWG NO.

	DRAWN BY	DATE	DWG NO.

Obtain an edge view of surface DEF. Also show point P in all views. Label the perpendicular distance from P to surface DEF, "d," in the appropriate view.

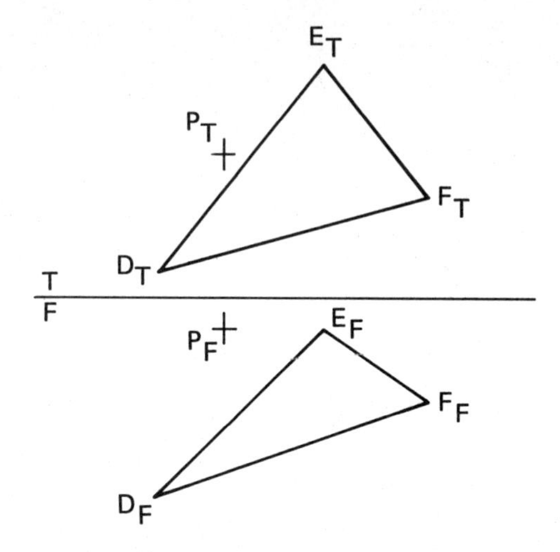

Title: Edge View of a Plane

A

Determine the point of cable RS and also show cable PQ in all views. Label the perpendicular distance, "d," between the cables, in the appropriate view.

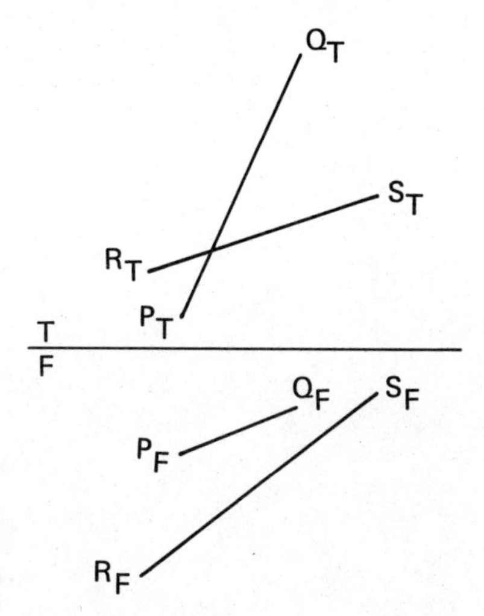

Title: Point View of a Line

B

DESCRIPTIVE GEOMETRY	DRAWN BY	DATE	DWG NO.

DRAWN BY
DATE
DWG NO.

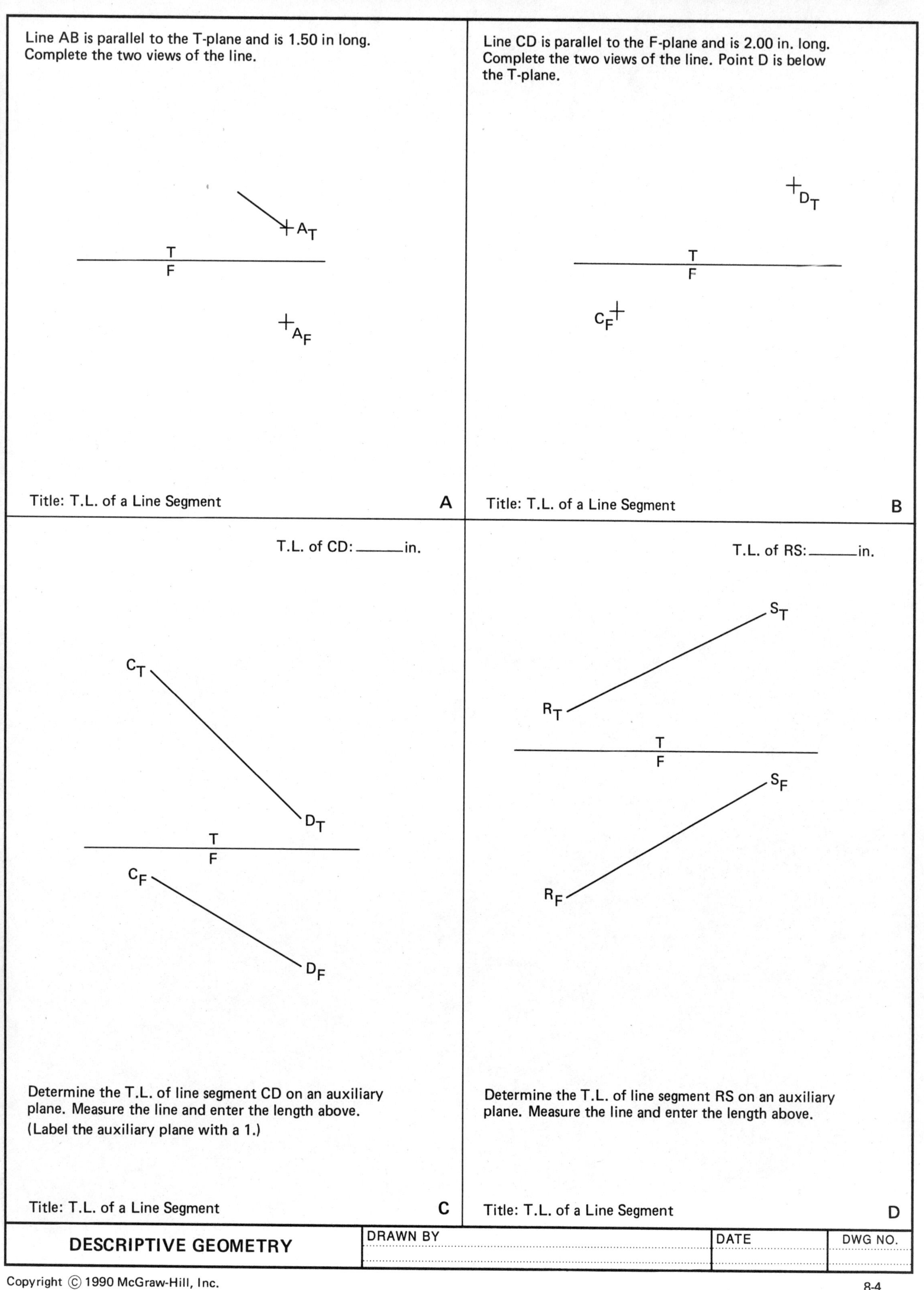

Line AB is parallel to the T-plane and is 1.50 in long. Complete the two views of the line.
A
T
T
F
A
F
Title: T.L. of a Line Segment
A
Line CD is parallel to the F-plane and is 2.00 in. long. Complete the two views of the line. Point D is below the T-plane.
D
T
T
F
C
F
Title: T.L. of a Line Segment
B
T.L. of CD: ______ in.
C
T
D
T
T
F
C
F
D
F
Determine the T.L. of line segment CD on an auxiliary plane. Measure the line and enter the length above. (Label the auxiliary plane with a 1.)
Title: T.L. of a Line Segment
C
T.L. of RS: ______ in.
S
T
R
T
T
F
S
F
R
F
Determine the T.L. of line segment RS on an auxiliary plane. Measure the line and enter the length above.
Title: T.L. of a Line Segment
D
DESCRIPTIVE GEOMETRY
DRAWN BY
DATE
DWG NO.

DRAWN BY
DATE
DWG NO.

Determine the true length of edge AB of the *Anchor Block*.
Measure the true length. Scale: 1″ = 1′–0.

T.L. of CD: ______ ft ______ in.

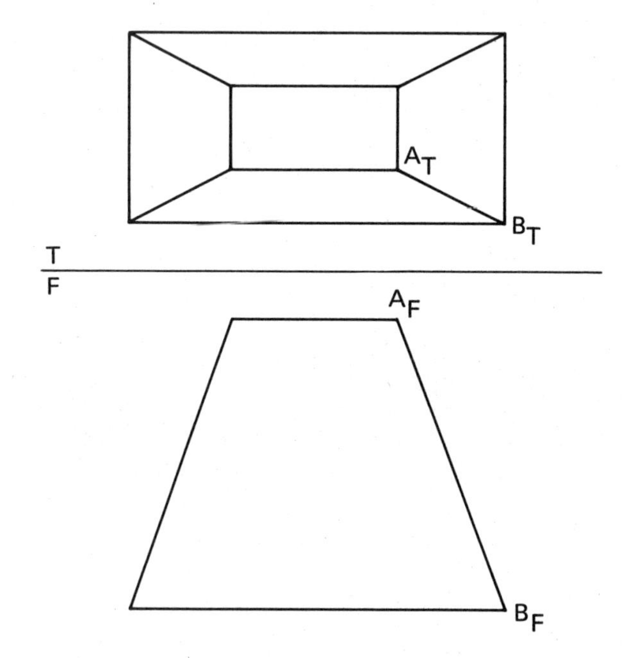

Title: True Length

A

Determine the true length of edge CD of the *Abutment.*
Scale: 1″ = 1′–0.

T.L of AB: ______ ft ______ in.

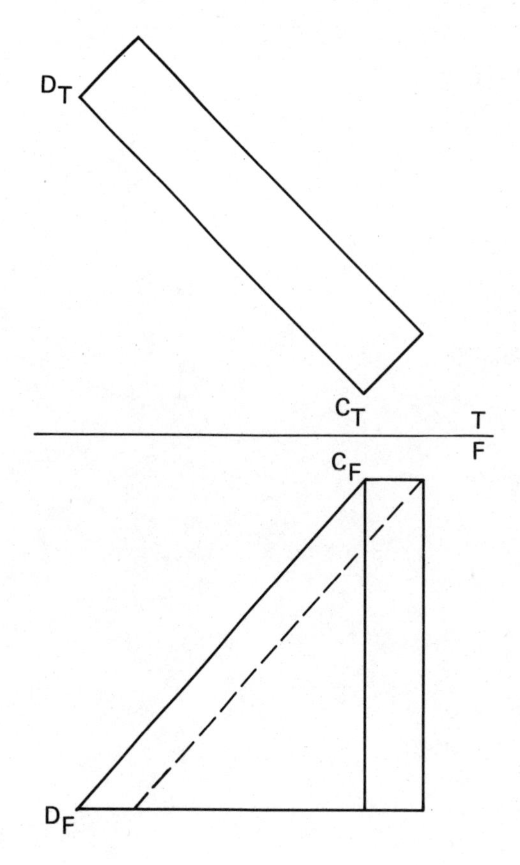

Title: True Length

B

DESCRIPTIVE GEOMETRY	DRAWN BY	DATE	DWG NO.

DRAWN BY
DATE
DWG NO.

The edge view of a plane can be seen on a view that shows the point view of some line in the plane.

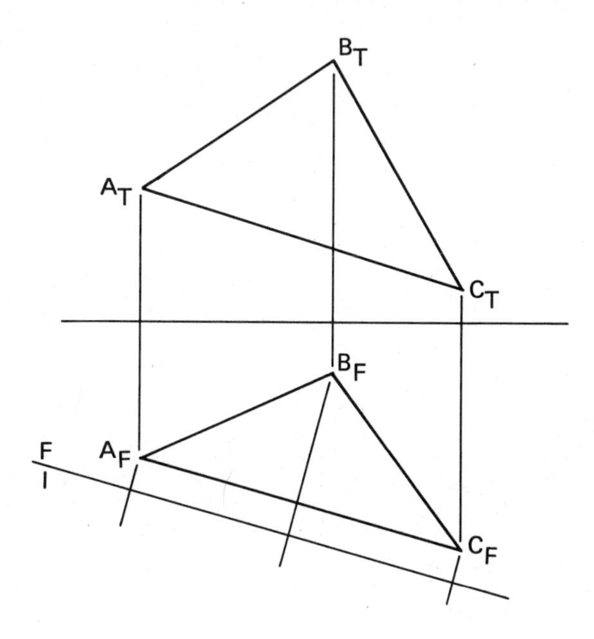

Determine the edge view of ABC.

Title: Edge View of a Plane

A

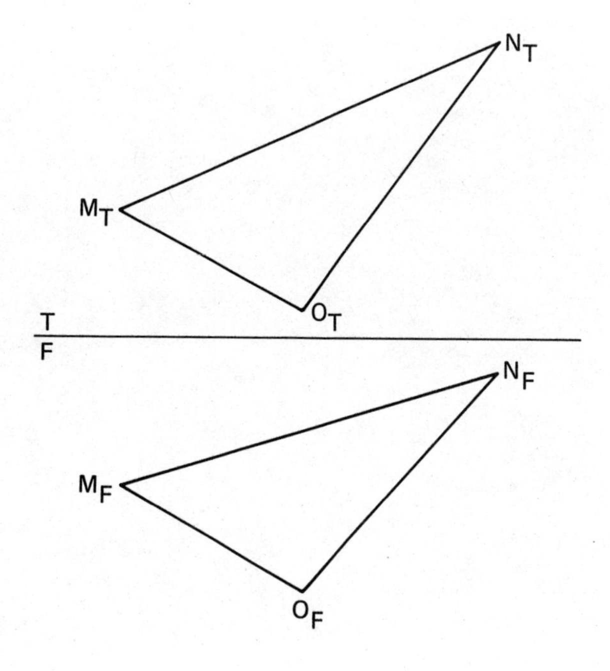

Determine the edge view of triangle MNO.

Title: Edge View of a Plane

B

DESCRIPTIVE GEOMETRY	DRAWN BY	DATE	DWG NO.

DRAWN BY
DATE
DWG NO.

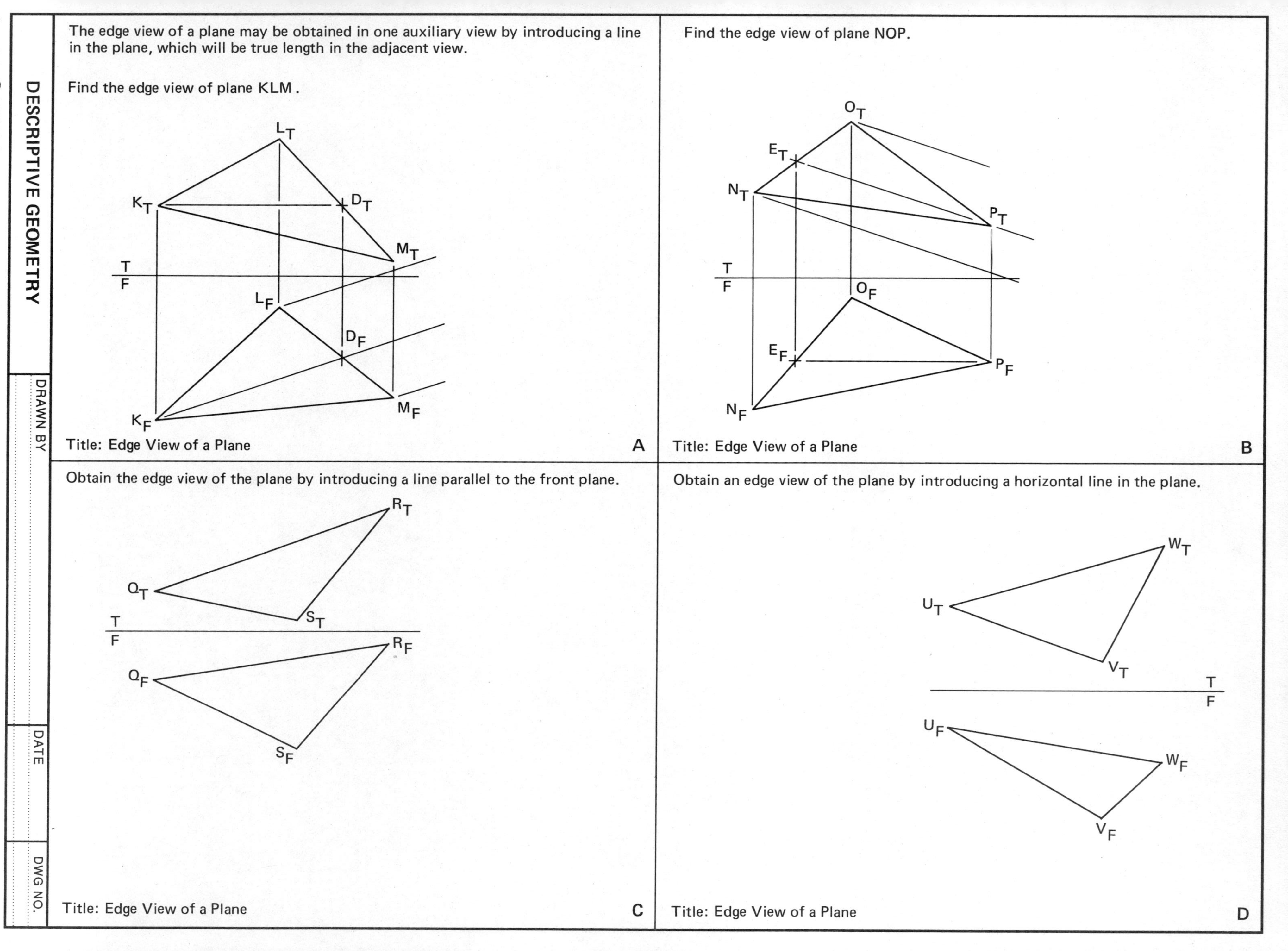

The edge view of a plane may be obtained in one auxiliary view by introducing a line in the plane, which will be true length in the adjacent view.
Find the edge view of plane KLM.
Title: Edge View of a Plane
A
Find the edge view of plane NOP.
Title: Edge View of a Plane
B
Obtain the edge view of the plane by introducing a line parallel to the front plane.
Title: Edge View of a Plane
C
Obtain an edge view of the plane by introducing a horizontal line in the plane.
Title: Edge View of a Plane
D
DESCRIPTIVE GEOMETRY
DRAWN BY
DATE
DWG NO.

DRAWN BY
DATE
DWG NO.

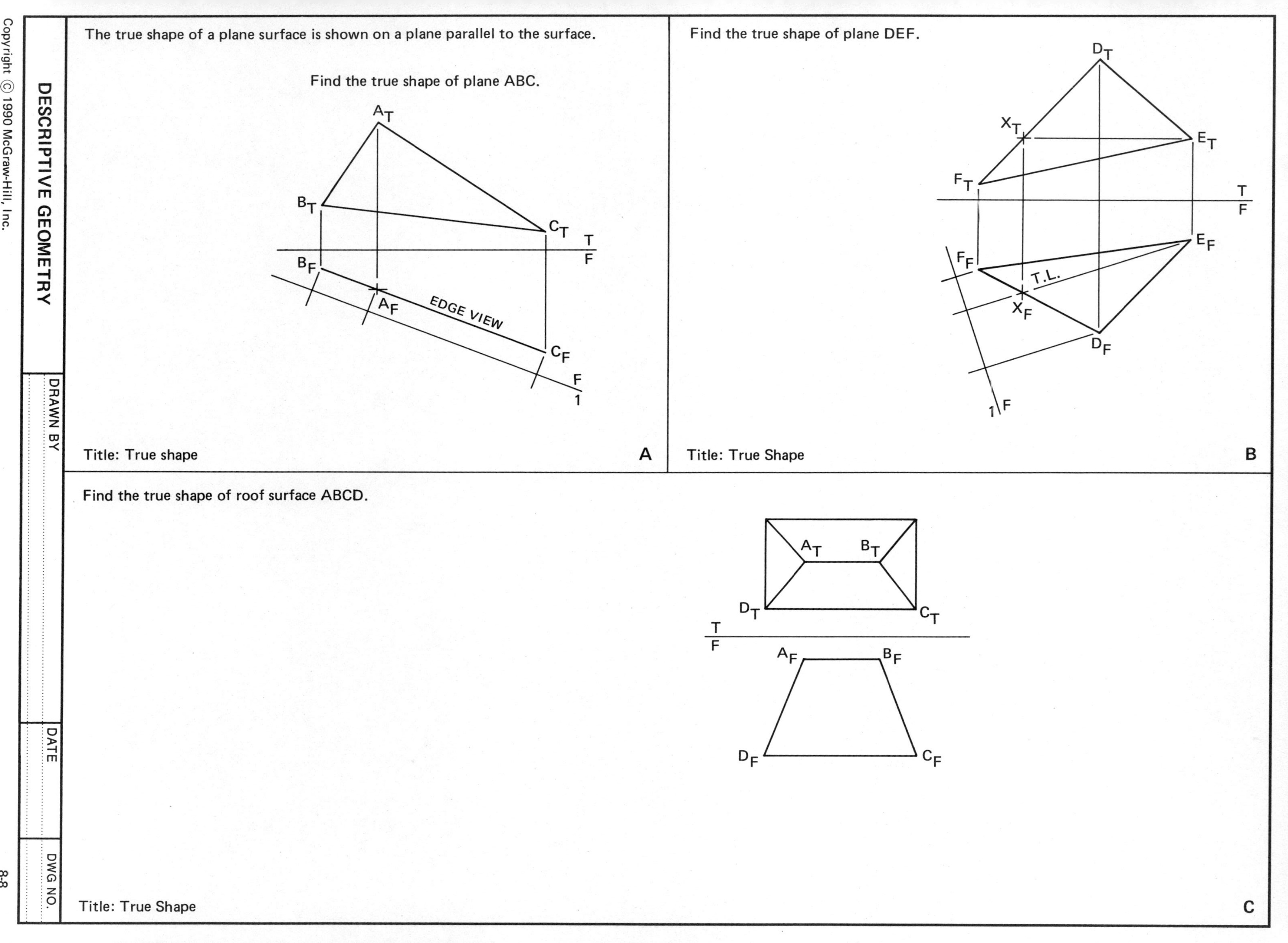
The true shape of a plane surface is shown on a plane parallel to the surface.
Find the true shape of plane ABC.
A_T
B_T
C_T
T
F
B_F
A_F
EDGE VIEW
C_F
F
1
Title: True shape
A
Find the true shape of plane DEF.
D_T
X_T
E_T
F_T
T
F
E_F
F_F
T.L.
X_F
D_F
1
F
Title: True Shape
B
Find the true shape of roof surface ABCD.
A_T
B_T
D_T
C_T
T
F
A_F
B_F
D_F
C_F
Title: True Shape
C
DESCRIPTIVE GEOMETRY
DRAWN BY
DATE
DWG NO.

DRAWN BY
DATE
DWG NO.

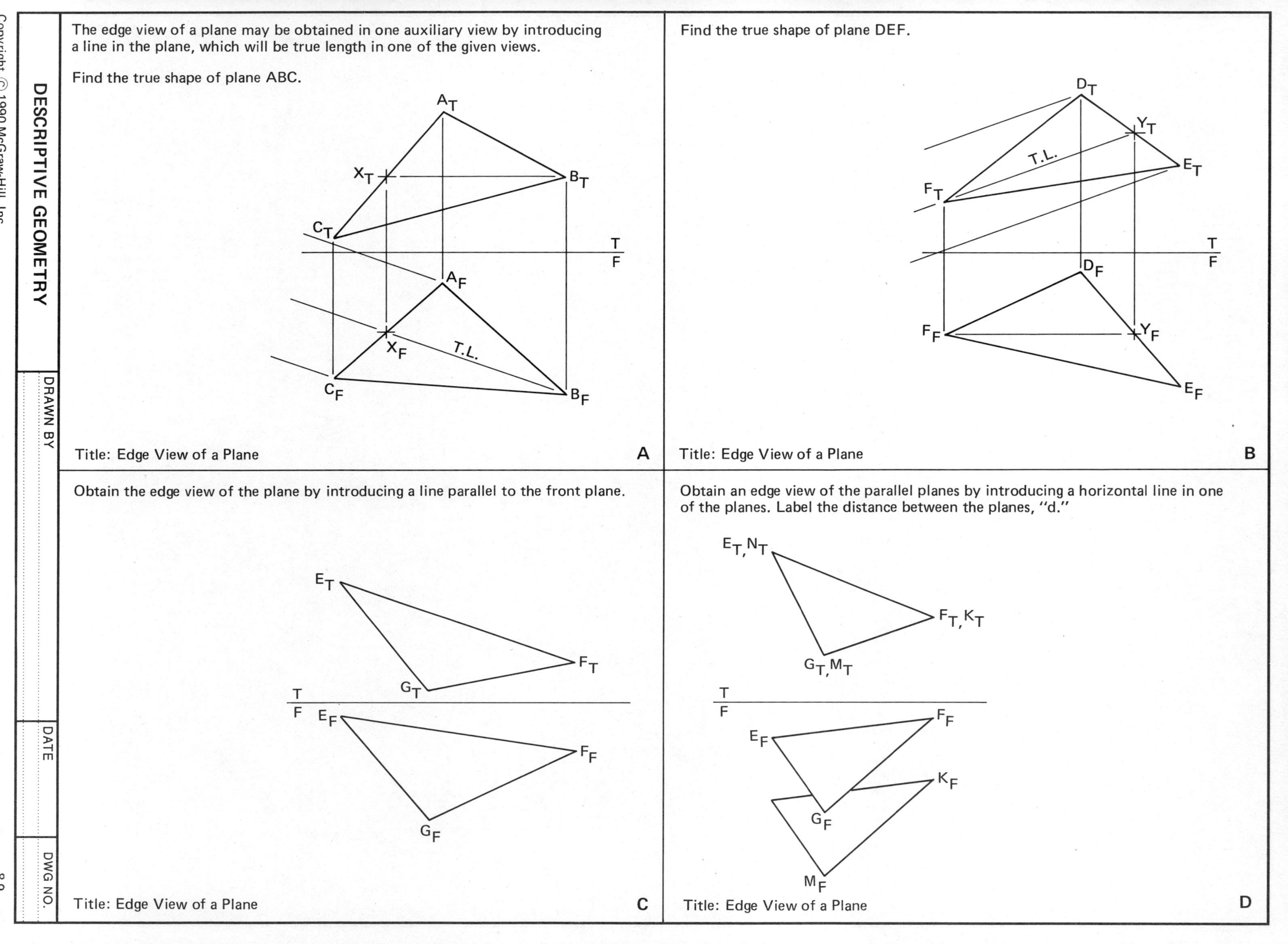

The edge view of a plane may be obtained in one auxiliary view by introducing a line in the plane, which will be true length in one of the given views.
Find the true shape of plane ABC.
A_T
X_T
B_T
C_T
T
F
A_F
X_F
T.L.
C_F
B_F
Title: Edge View of a Plane
A
Find the true shape of plane DEF.
D_T
Y_T
T.L.
E_T
F_T
T
F
D_F
F_F
Y_F
E_F
Title: Edge View of a Plane
B
Obtain the edge view of the plane by introducing a line parallel to the front plane.
E_T
F_T
G_T
T
F
E_F
F_F
G_F
Title: Edge View of a Plane
C
Obtain an edge view of the parallel planes by introducing a horizontal line in one of the planes. Label the distance between the planes, "d."
E_T, N_T
F_T, K_T
G_T, M_T
T
F
F_F
E_F
K_F
G_F
M_F
Title: Edge View of a Plane
D
DESCRIPTIVE GEOMETRY
DRAWN BY
DATE
DWG NO.

DRAWN BY
DATE
DWG NO.

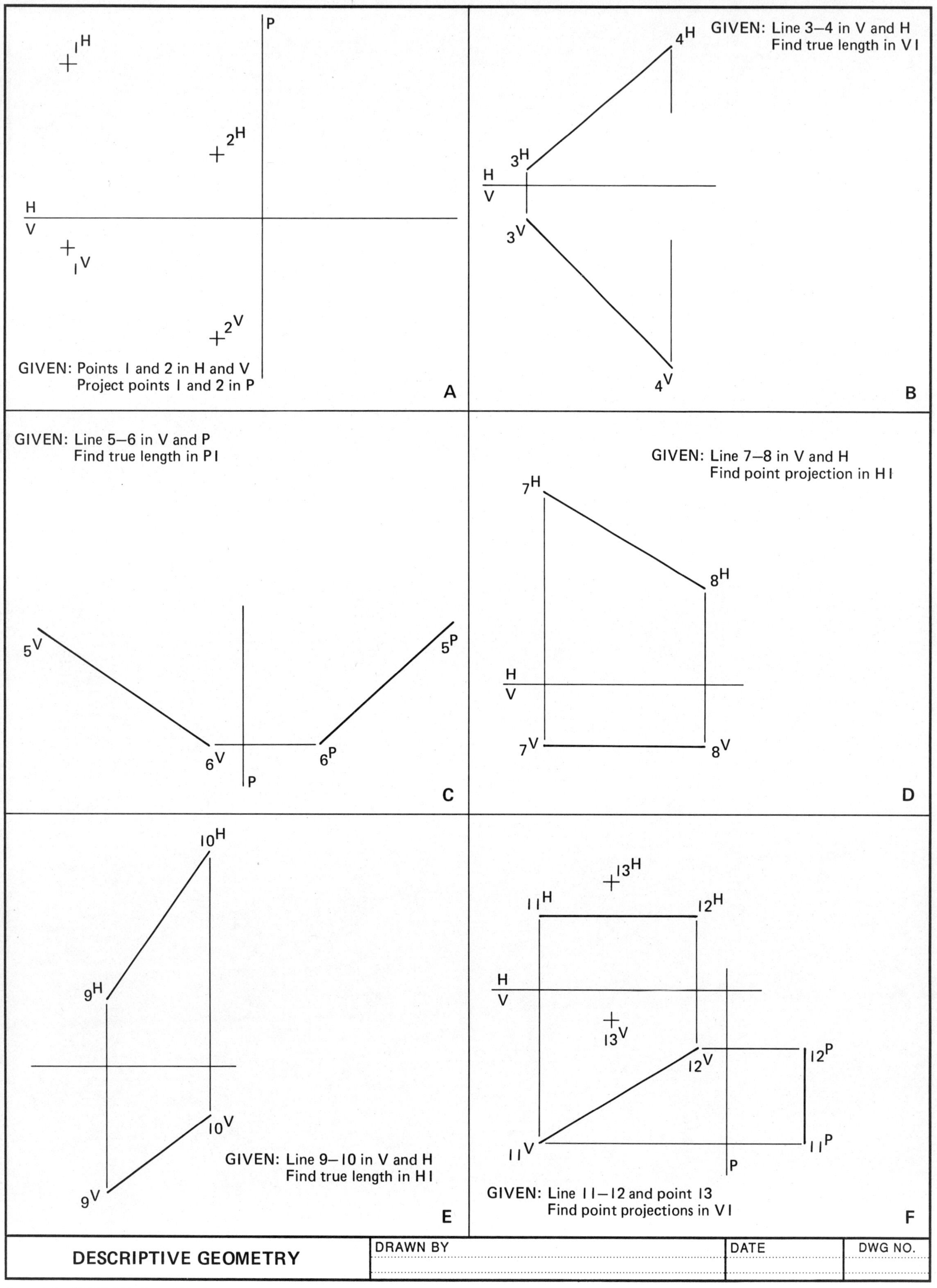
GIVEN: Points 1 and 2 in H and V
Project points 1 and 2 in P
A
GIVEN: Line 3–4 in V and H
Find true length in V1
B
GIVEN: Line 5–6 in V and P
Find true length in P1
C
GIVEN: Line 7–8 in V and H
Find point projection in H1
D
GIVEN: Line 9–10 in V and H
Find true length in H1
E
GIVEN: Line 11–12 and point 13
Find point projections in V1
F
DESCRIPTIVE GEOMETRY
DRAWN BY
DATE
DWG NO.

	DRAWN BY	DATE	DWG NO.

A TORUS THAT TWISTS

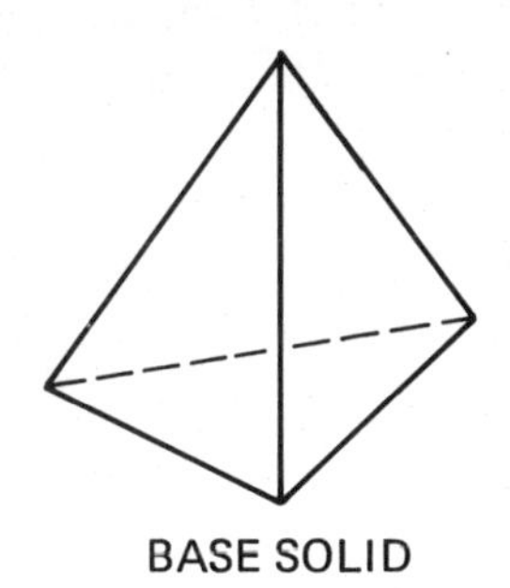

BASE SOLID

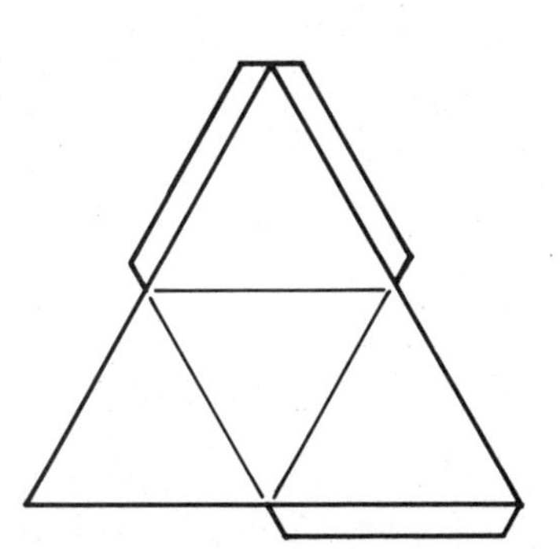

TETRAHEDRON PATTERN

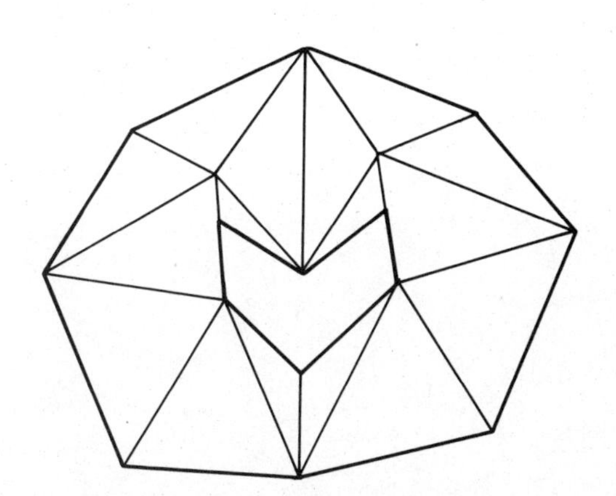

A RING OF TETRAHEDRONS

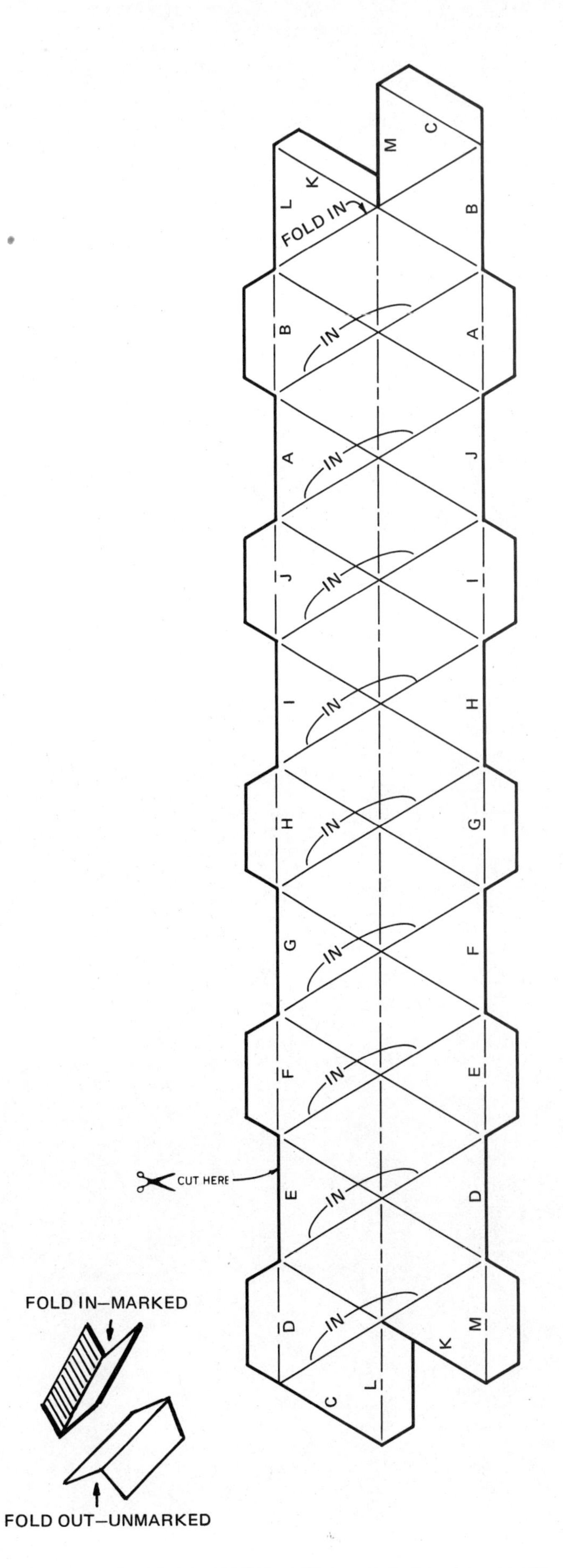

PROBLEM:

Cut out the pattern and fold as shown.
Connect the letters and tape the corners
to form a *Torus* that twists.

DESCRIPTIVE GEOMETRY	DRAWN BY	DATE	DWG NO.

	DRAWN BY	DATE	DWG NO.

Complete the front and top views of the *Swivel Hanger.* Add a right-side view in full section.

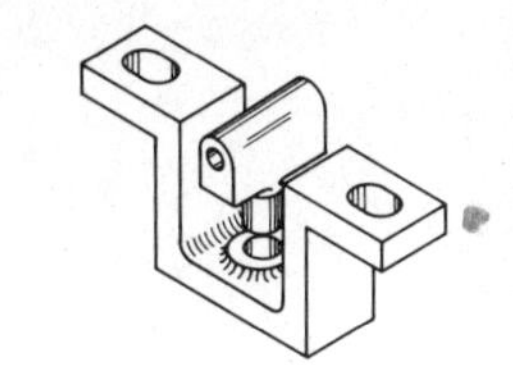

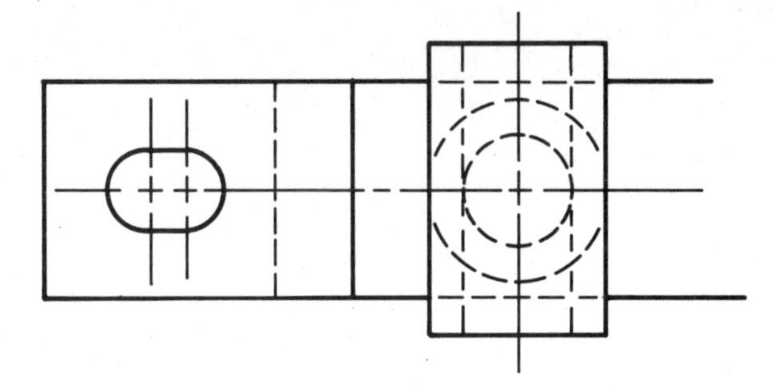

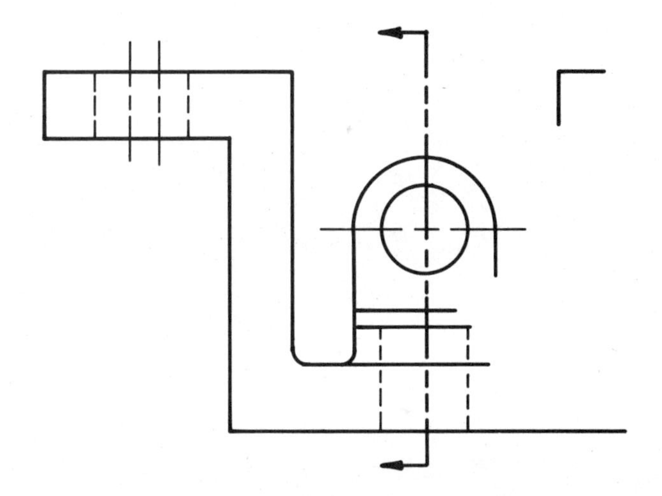

A

Draw a half section and a full section of the *Cushion Wheel* in the space provided at the right. Study the given views carefully before beginning. Section lining should correspond to the materials noted. Some hidden lines were omitted for clarity.

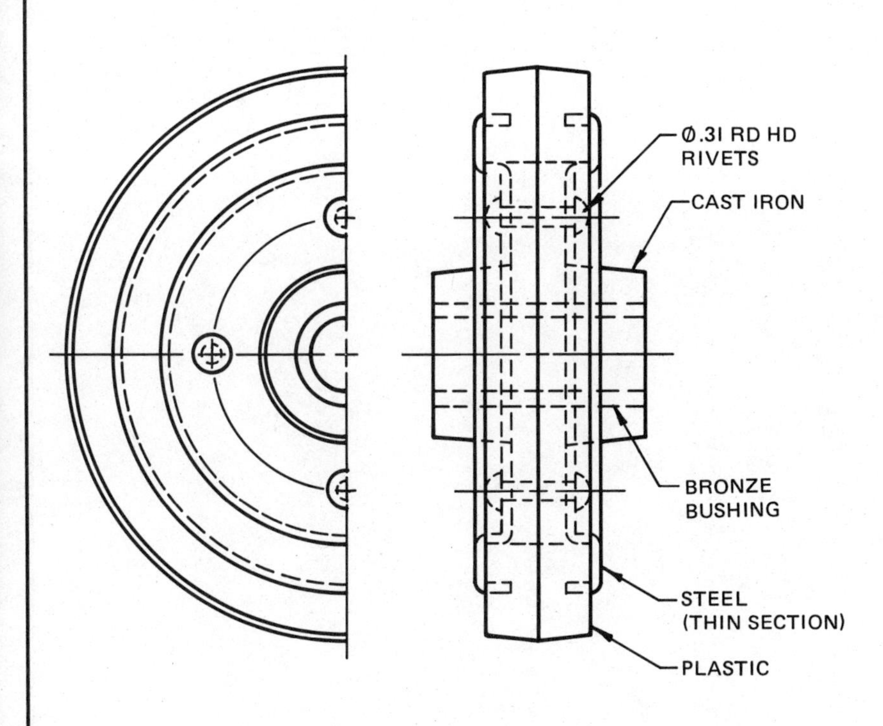

HALF SECTION

FULL SECTION

B

SECTIONAL VIEWS AND CONVENTIONS	DRAWN BY	DATE	DWG NO.

	DRAWN BY	DATE	DWG NO.

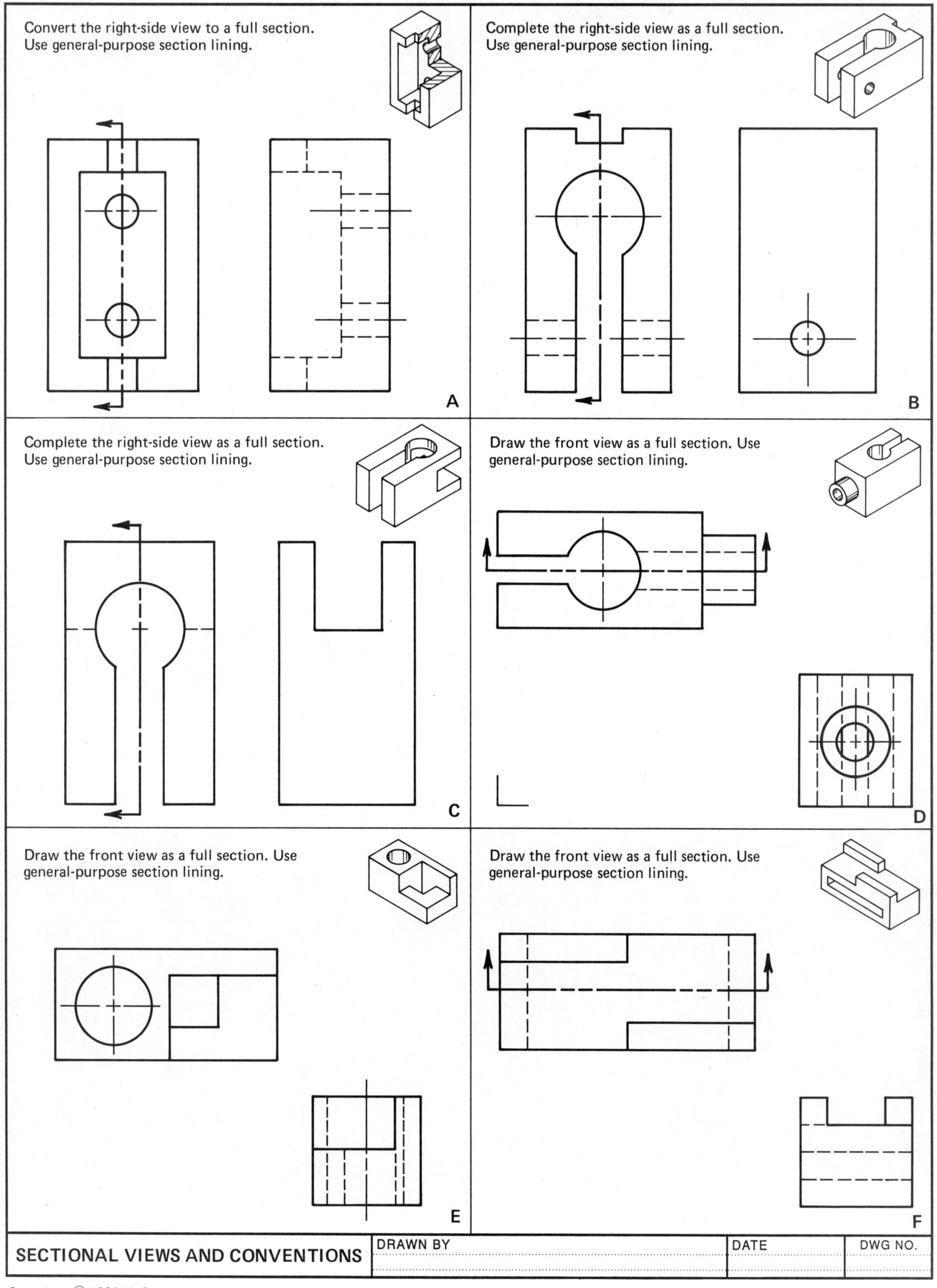
Convert the right-side view to a full section. Use general-purpose section lining.
A
Complete the right-side view as a full section. Use general-purpose section lining.
B
Complete the right-side view as a full section. Use general-purpose section lining.
C
Draw the front view as a full section. Use general-purpose section lining.
D
Draw the front view as a full section. Use general-purpose section lining.
E
Draw the front view as a full section. Use general-purpose section lining.
F
SECTIONAL VIEWS AND CONVENTIONS
DRAWN BY
DATE
DWG NO.

DRAWN BY
DATE
DWG NO.

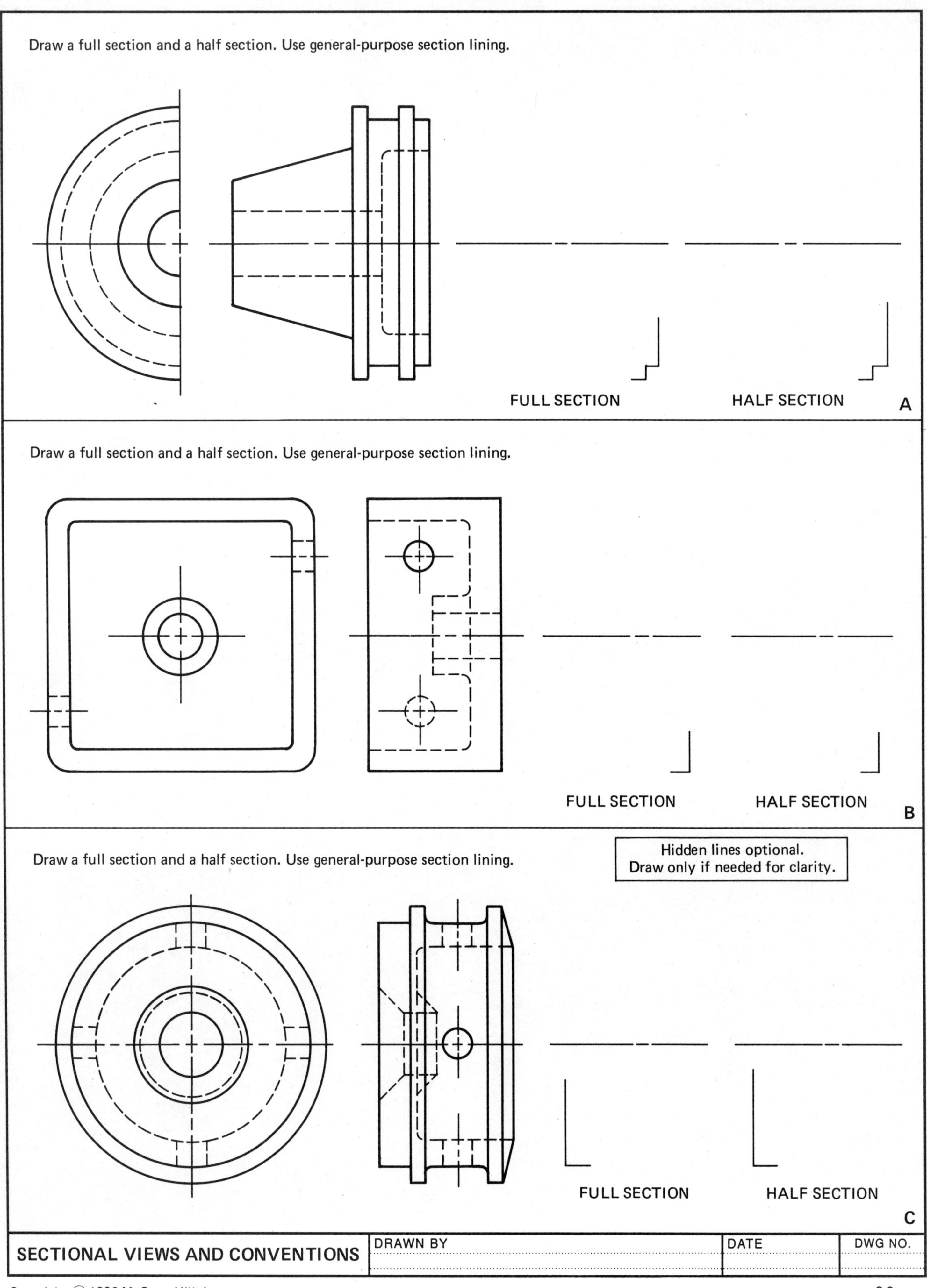
Draw a full section and a half section. Use general-purpose section lining.
FULL SECTION
HALF SECTION
A
Draw a full section and a half section. Use general-purpose section lining.
FULL SECTION
HALF SECTION
B
Draw a full section and a half section. Use general-purpose section lining.
Hidden lines optional.
Draw only if needed for clarity.
FULL SECTION
HALF SECTION
C
SECTIONAL VIEWS AND CONVENTIONS
DRAWN BY
DATE
DWG NO.

DRAWN BY
DATE
DWG NO.

Draw an offset section passing through A, B, and C. Add the cutting-plane line to the top view. Use general-purpose section lining.

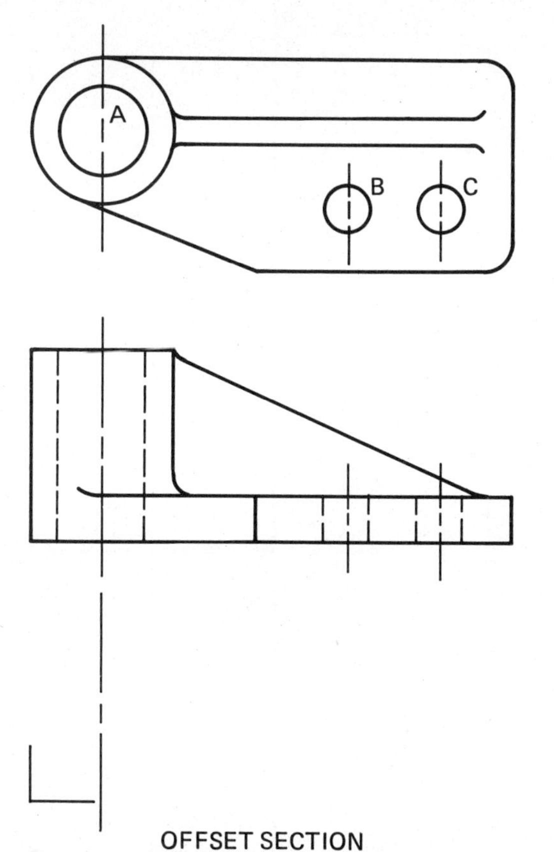

OFFSET SECTION

A

Draw a full section. Use general-purpose section lining.

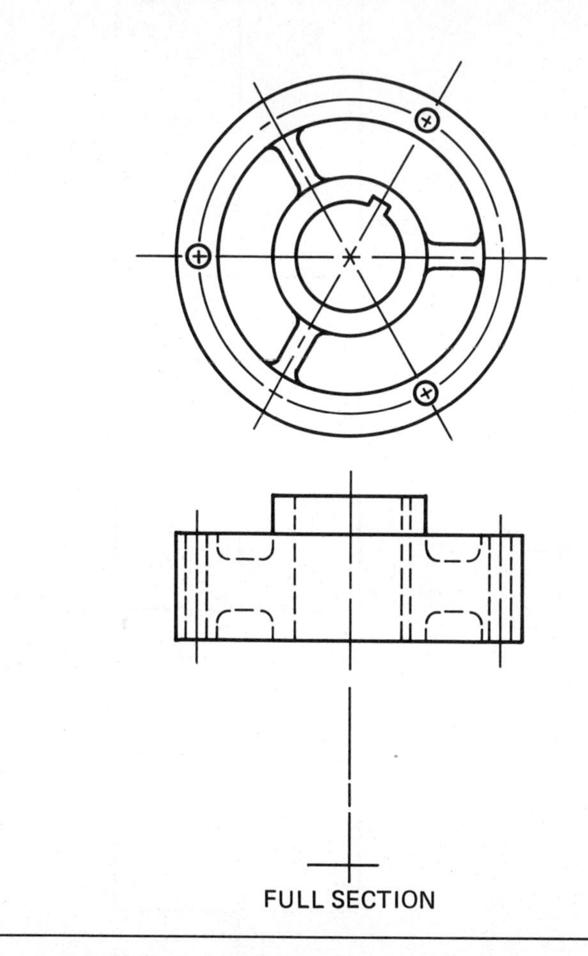

FULL SECTION

B

Draw removed sections A—A and B—B. Use general-purpose section lining.

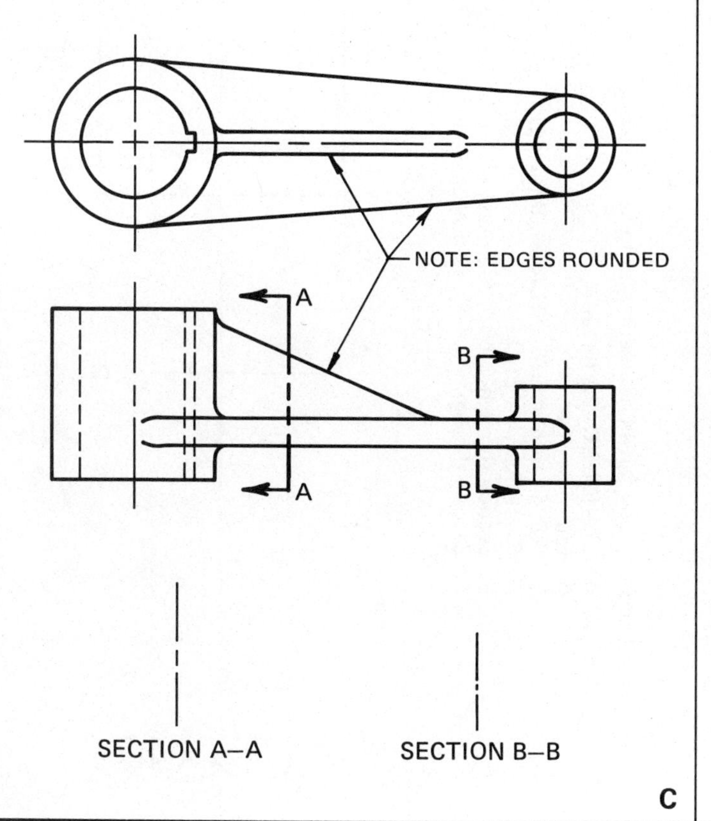

SECTION A—A

SECTION B—B

C

Draw a revolved section on centerline A—A. Complete the view at the bottom by showing a broken-out section. Use general-purpose section lining.

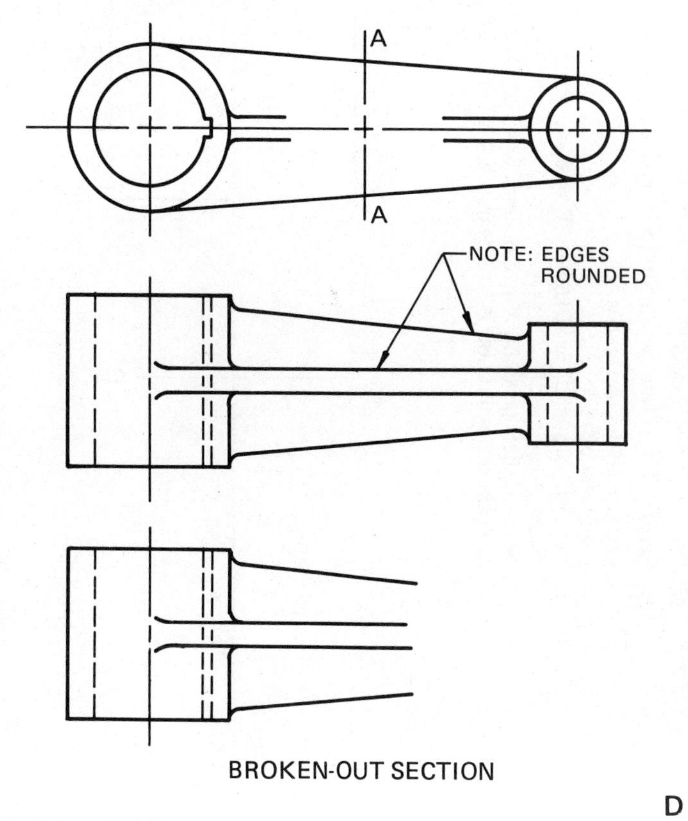

BROKEN-OUT SECTION

D

SECTIONAL VIEWS AND CONVENTIONS	DRAWN BY	DATE	DWG NO.

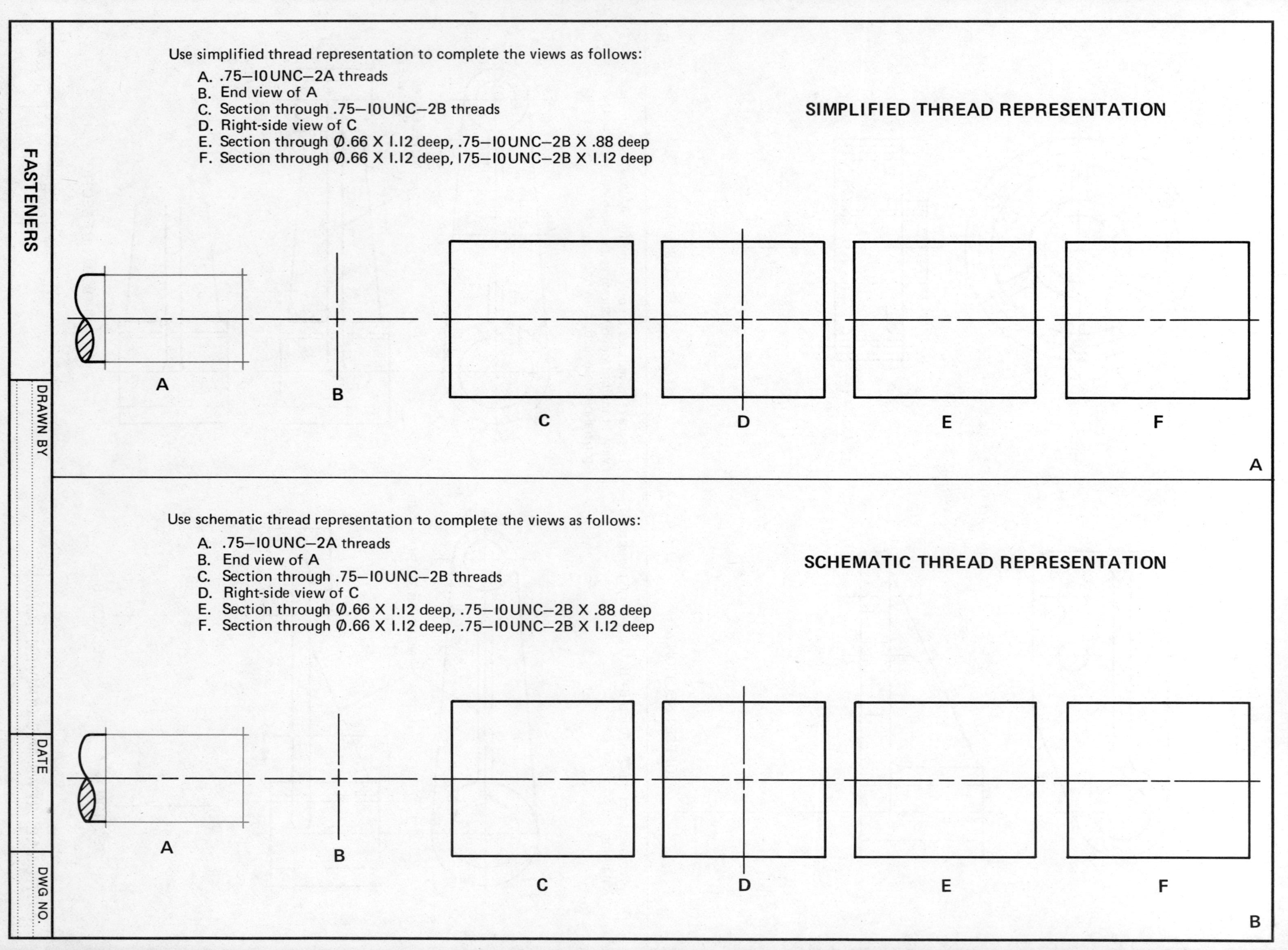

Use simplified thread representation to complete the views as follows:
A. .75–10UNC–2A threads
B. End view of A
C. Section through .75–10UNC–2B threads
D. Right-side view of C
E. Section through Ø.66 X 1.12 deep, .75–10UNC–2B X .88 deep
F. Section through Ø.66 X 1.12 deep, I75–10UNC–2B X 1.12 deep
SIMPLIFIED THREAD REPRESENTATION
A
B
C
D
E
F
A
Use schematic thread representation to complete the views as follows:
A. .75–10UNC–2A threads
B. End view of A
C. Section through .75–10UNC–2B threads
D. Right-side view of C
E. Section through Ø.66 X 1.12 deep, .75–10UNC–2B X .88 deep
F. Section through Ø.66 X 1.12 deep, .75–10UNC–2B X 1.12 deep
SCHEMATIC THREAD REPRESENTATION
A
B
C
D
E
F
B
FASTENERS
DRAWN BY
DATE
DWG NO.

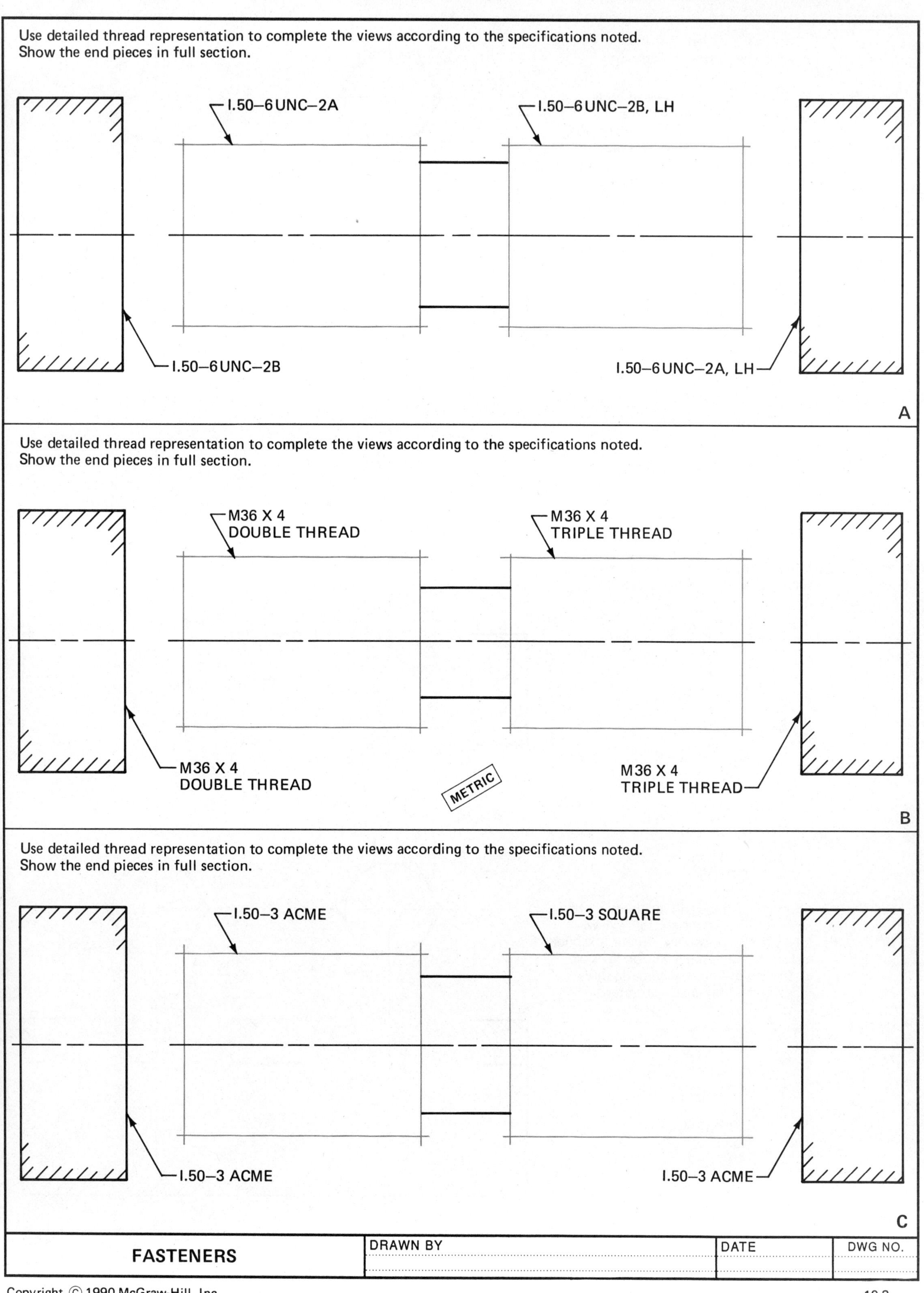
Use detailed thread representation to complete the views according to the specifications noted.
Show the end pieces in full section.
I.50–6 UNC–2A
I.50–6 UNC–2B, LH
I.50–6 UNC–2B
I.50–6 UNC–2A, LH
A
Use detailed thread representation to complete the views according to the specifications noted.
Show the end pieces in full section.
M36 X 4
DOUBLE THREAD
M36 X 4
TRIPLE THREAD
M36 X 4
DOUBLE THREAD
METRIC
M36 X 4
TRIPLE THREAD
B
Use detailed thread representation to complete the views according to the specifications noted.
Show the end pieces in full section.
I.50–3 ACME
I.50–3 SQUARE
I.50–3 ACME
I.50–3 ACME
C
FASTENERS
DRAWN BY
DATE
DWG NO.

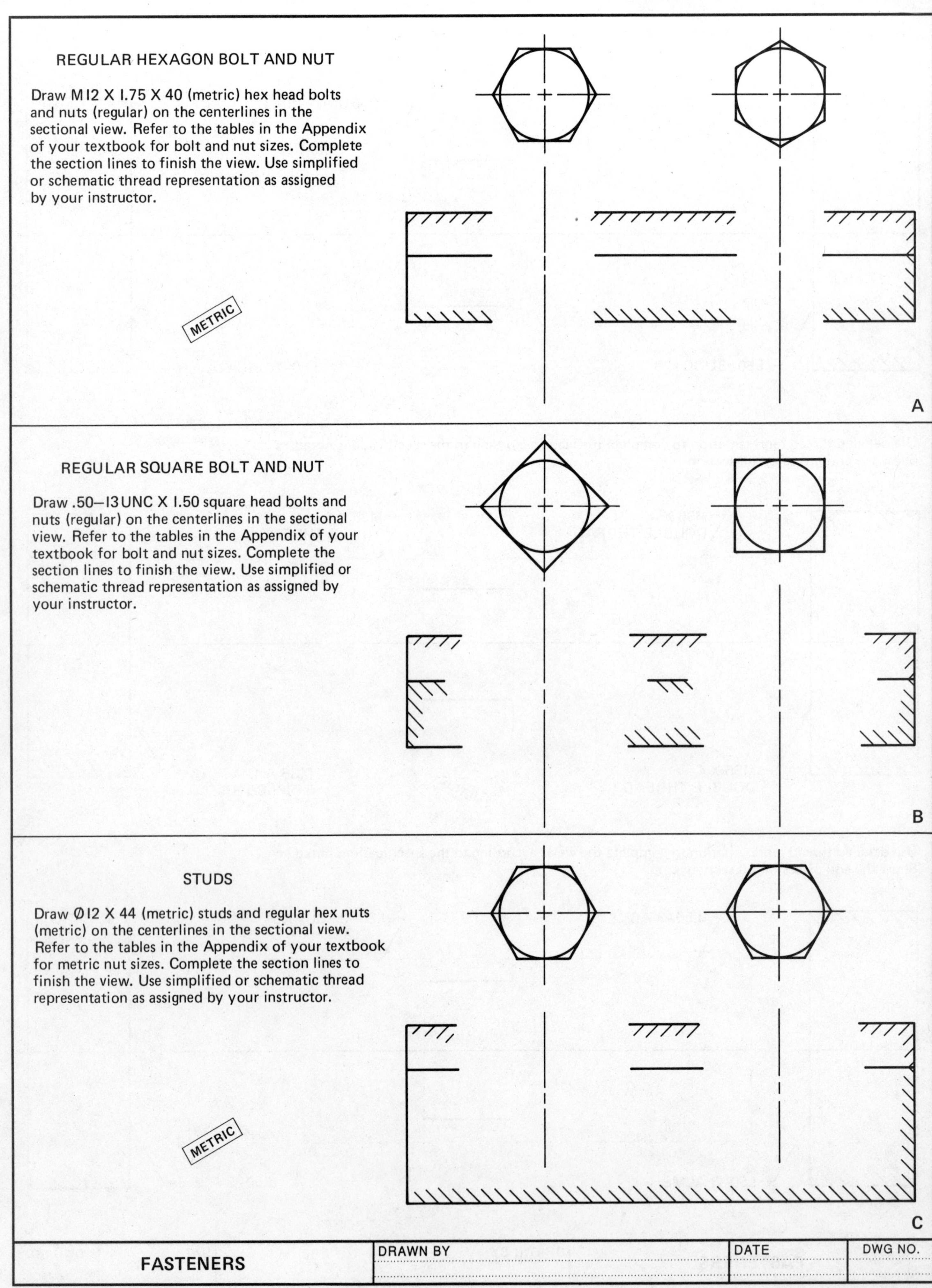
REGULAR HEXAGON BOLT AND NUT
Draw M12 X 1.75 X 40 (metric) hex head bolts and nuts (regular) on the centerlines in the sectional view. Refer to the tables in the Appendix of your textbook for bolt and nut sizes. Complete the section lines to finish the view. Use simplified or schematic thread representation as assigned by your instructor.
METRIC
A
REGULAR SQUARE BOLT AND NUT
Draw .50–13 UNC X 1.50 square head bolts and nuts (regular) on the centerlines in the sectional view. Refer to the tables in the Appendix of your textbook for bolt and nut sizes. Complete the section lines to finish the view. Use simplified or schematic thread representation as assigned by your instructor.
B
STUDS
Draw Ø12 X 44 (metric) studs and regular hex nuts (metric) on the centerlines in the sectional view. Refer to the tables in the Appendix of your textbook for metric nut sizes. Complete the section lines to finish the view. Use simplified or schematic thread representation as assigned by your instructor.
METRIC
C
FASTENERS
DRAWN BY
DATE
DWG NO.

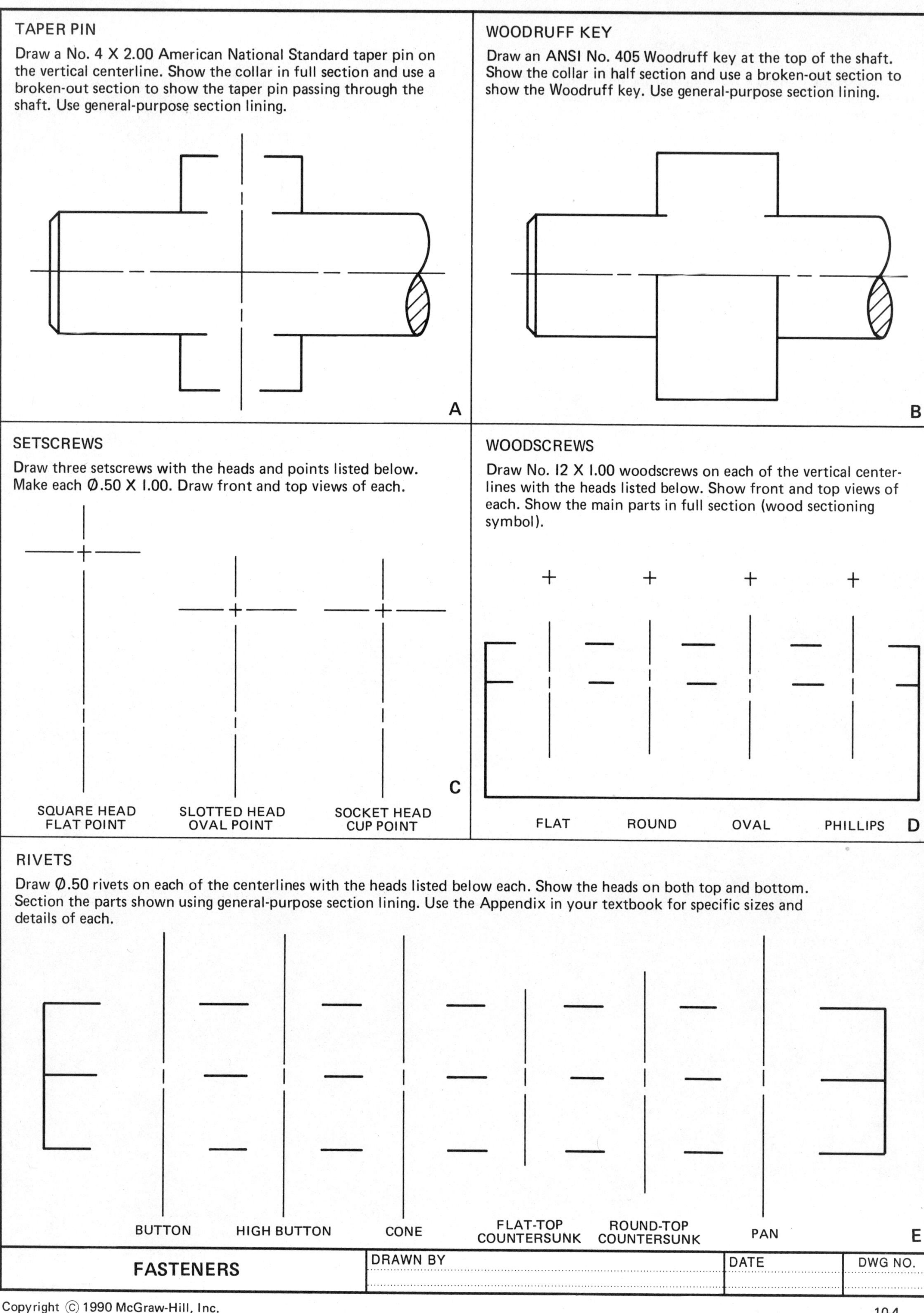
TAPER PIN
Draw a No. 4 X 2.00 American National Standard taper pin on the vertical centerline. Show the collar in full section and use a broken-out section to show the taper pin passing through the shaft. Use general-purpose section lining.
A
WOODRUFF KEY
Draw an ANSI No. 405 Woodruff key at the top of the shaft. Show the collar in half section and use a broken-out section to show the Woodruff key. Use general-purpose section lining.
B
SETSCREWS
Draw three setscrews with the heads and points listed below. Make each Ø.50 X 1.00. Draw front and top views of each.
SQUARE HEAD FLAT POINT
SLOTTED HEAD OVAL POINT
SOCKET HEAD CUP POINT
C
WOODSCREWS
Draw No. 12 X 1.00 woodscrews on each of the vertical centerlines with the heads listed below. Show front and top views of each. Show the main parts in full section (wood sectioning symbol).
FLAT
ROUND
OVAL
PHILLIPS
D
RIVETS
Draw Ø.50 rivets on each of the centerlines with the heads listed below each. Show the heads on both top and bottom. Section the parts shown using general-purpose section lining. Use the Appendix in your textbook for specific sizes and details of each.
BUTTON
HIGH BUTTON
CONE
FLAT-TOP COUNTERSUNK
ROUND-TOP COUNTERSUNK
PAN
E
FASTENERS
DRAWN BY
DATE
DWG NO.

STOP CLAMP

ASSIGNMENT I: Make a working drawing of the stop clamp. Scale—optional. Dimension. Body—die-cast aluminum. Knurled screw—cold-rolled steel.

ASSIGNMENT 2: Make an assembly drawing of the stop clamp. Dimension if required by your instructor. Estimate any sizes not given.

NOTE: Dimensions may be converted from fractional-inch to decimal-inch or metric. Conversion charts are included in the Appendix of your textbook.

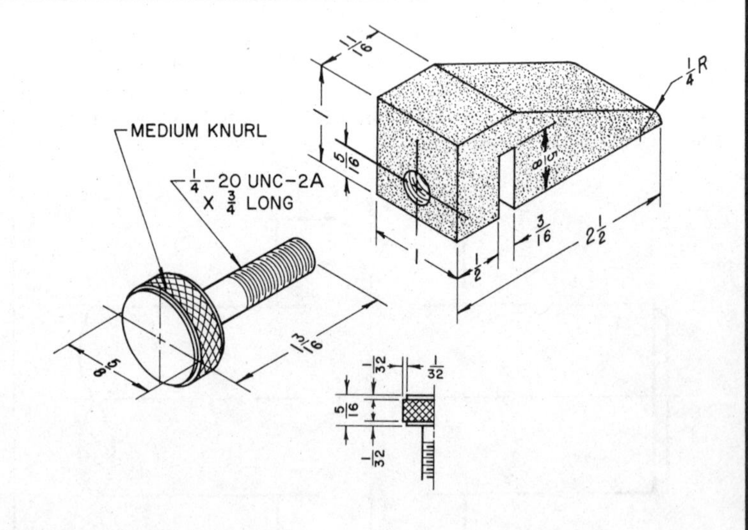

WORKING DRAWINGS	DRAWN BY	DATE	DWG NO.

END BASE

Make a working drawing of the *End Base*. Scale: optional. Dimension. Use partial and/or sectional views where needed. Material: cast iron. Estimate sizes not given.

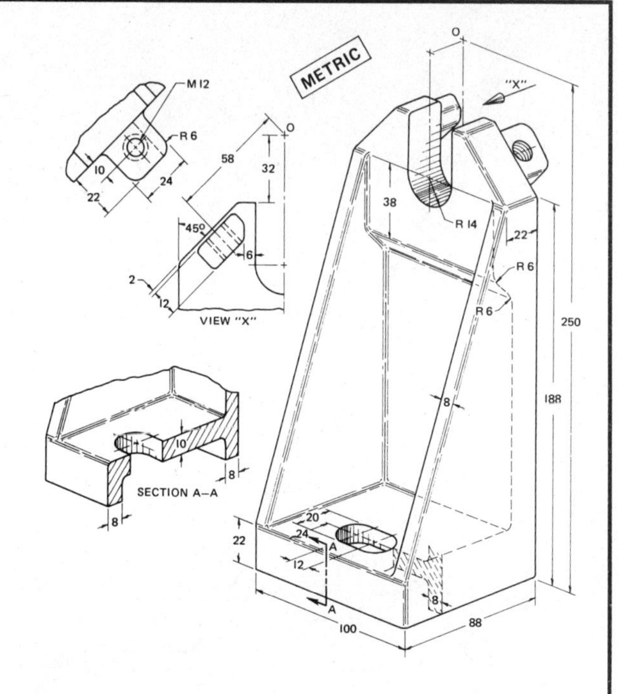

WORKING DRAWINGS	DRAWN BY	DATE	DWG NO.

MARKING GAGE

ASSIGNMENT 1: On drawing sheet 11-3A, make a working drawing of each part. Scale: optional. Dimension. Head: cast iron. Beam and knurled screw: CRS. Scriber: 90 point carbon steel. Face of head is to be machined.

ASSIGNMENT 2: (See drawing sheet 11-3B)

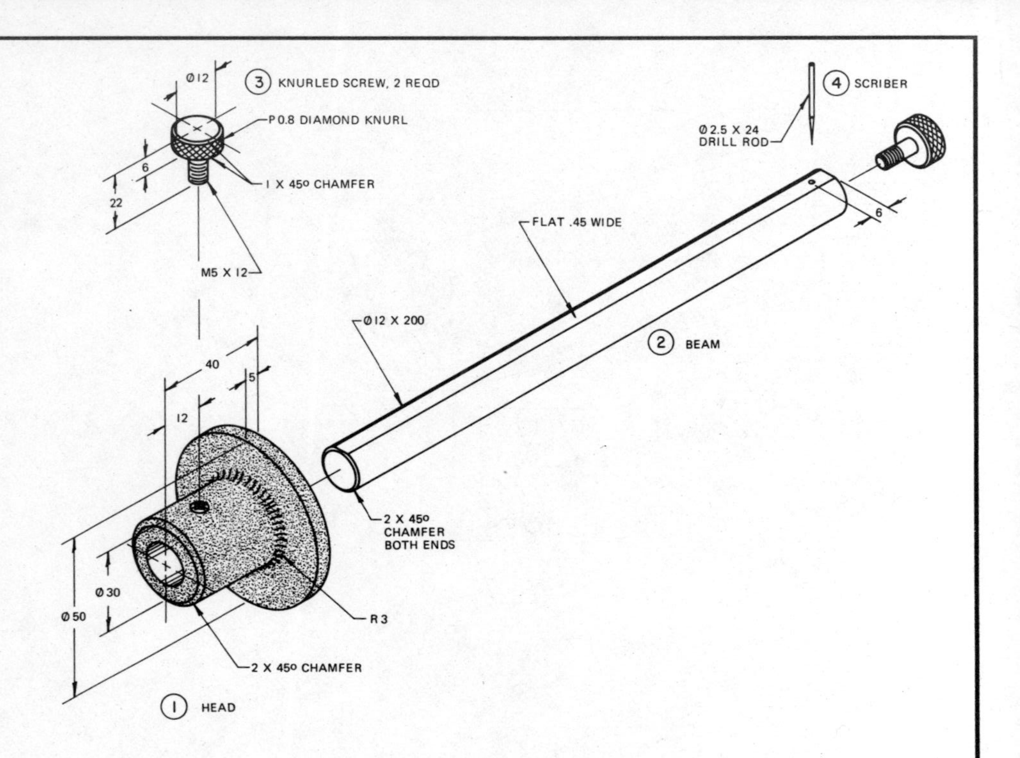

WORKING DRAWINGS	DRAWN BY	DATE	DWG NO.

MARKING GAGE
(ASSEMBLY DRAWING AND BILL OF MATERIAL)

ASSIGNMENT 2: Make an assembly drawing and bill of material for the *Marking Gage*. Include a bill of material if assigned. Use sectional views where appropriate. Do not dimension. Use simplified thread representation.

WORKING DRAWINGS	DRAWN BY	DATE	DWG NO.

FLANGE

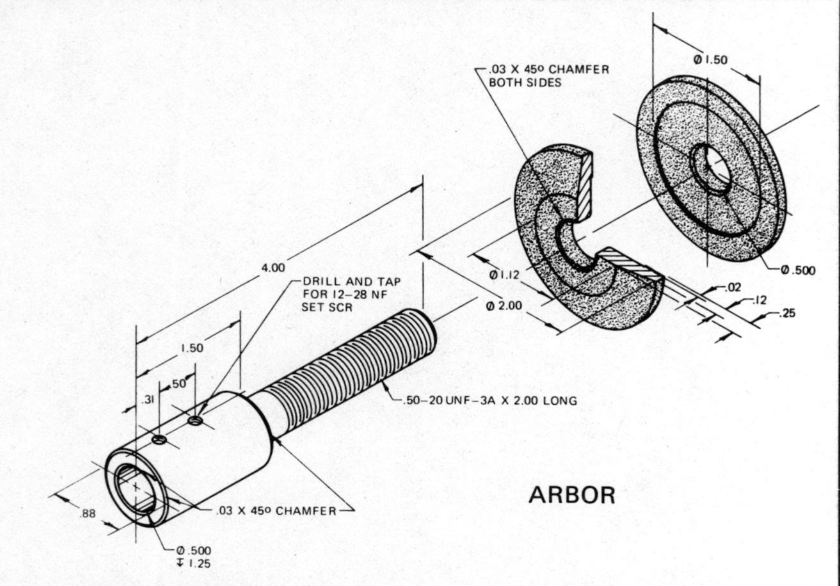

ARBOR

ASSIGNMENT 1: On drawing sheet 11-4A, make a working drawing of each part shown. Scale: full size or as assigned. Dimension. Flanges: die-cast aluminum. Shaft: cold-rolled steel. Section where appropriate for clarity. Use ANSI material symbols on sectional views.

ASSIGNMENT 2: (See drawing sheet 11-4B)

SHAFT

WORKING DRAWINGS	DRAWN BY	DATE	DWG NO.

ARBOR
(ASSEMBLY DRAWING AND BILL OF MATERIAL)

ASSIGNMENT 2: On drawing sheet 11-4B, make an assembly drawing of the *Arbor* with a Ø6.00 X 1.00 grinding wheel between the flanges. Include a bill of material if assigned. Be sure to include the setscrews. Use sectional views where appropriate. Draw all fasteners on the assembly view. Use simplified or schematic thread representation as assigned. Estimate all sizes and details not given.

NOTES:

BILL OF MATERIAL

DRAWING DRAWINGS	DRAWN BY	DATE	DWG NO.

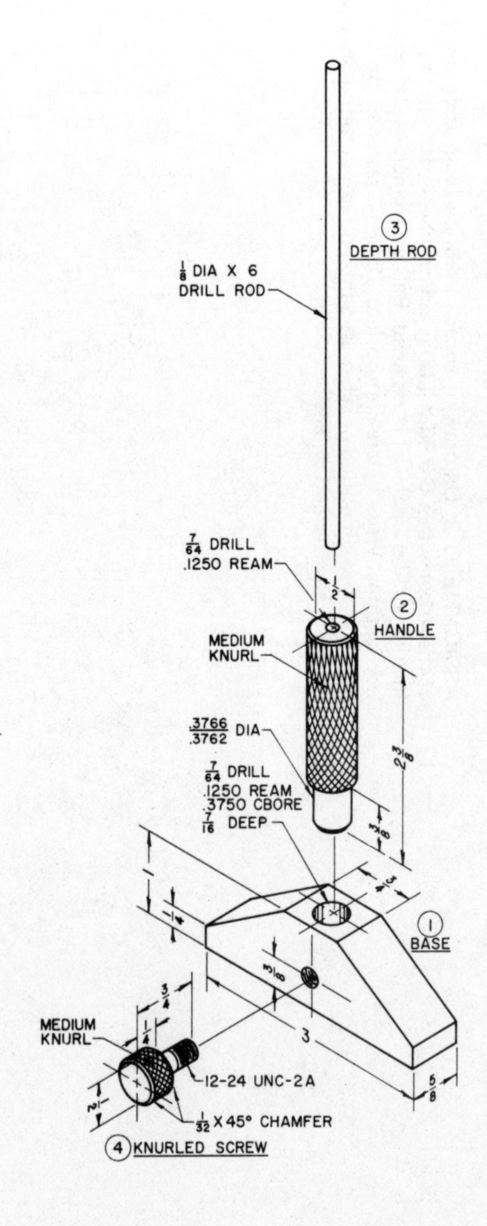

DEPTH GAGE

Make detail drawings of the parts for the *Depth Gage*. Scale: three-quarter size. Dimension all parts. All material CRS. All parts FAO. Since the drill rod (Part No. 3) is a stock item, it need not be detailed but must be included in a bill of material. Convert fractions to decimals or millimeters is assigned.

Make a two-view assembly drawing and bill of material for the *Depth Gage.*

NOTE: Drawing sheets 11-5A and 11-5B may be used for these assignments.

WORKING DRAWINGS	DRAWN BY	DATE	DWG NO.

Depth Gage
WORKING DRAWINGS
DRAWN BY
DATE
DWG NO.

TRAMMEL

METRIC

MEDIUM KNURL
Ø14
0.8
0.8
4
22
R 4
M6
12 LONG
Ø16
Ø8 X 600
CRS ROD WITH
5 WIDE FLAT
10
20
8
R 4
60
Ø9.5
M5
12 LONG
Ø8
MEDIUM KNURL
12
8
54
28
Ø 2.5

ASSIGNMENT 1: Make a detail detail drawing of each part shown. Scale: full size. Dimension. Specify "2 REQD" for the body, point, and knurled screw. Body, beam, and knurled screw CRS. The point is 90 point carbon steel and is to be heat treated after machining. Use sectional views where appropriate. For example, use a conventional break to shorten the length of the drawing of the beam.

KNURLED SCREW

BODY

BEAM

POINT

ASSEMBLY DRAWING OF TRAMMEL
SCALE: HALF SIZE

ASSIGNMENT 2: Make an assembly drawing and a bill of material. Scale: half size.

NOTE: The gray lines will help you to locate the center of each of the part drawings.

NOTES:

METRIC

BILL OF MATERIAL

WORKING DRAWINGS	DRAWN BY	DATE	DWG NO.

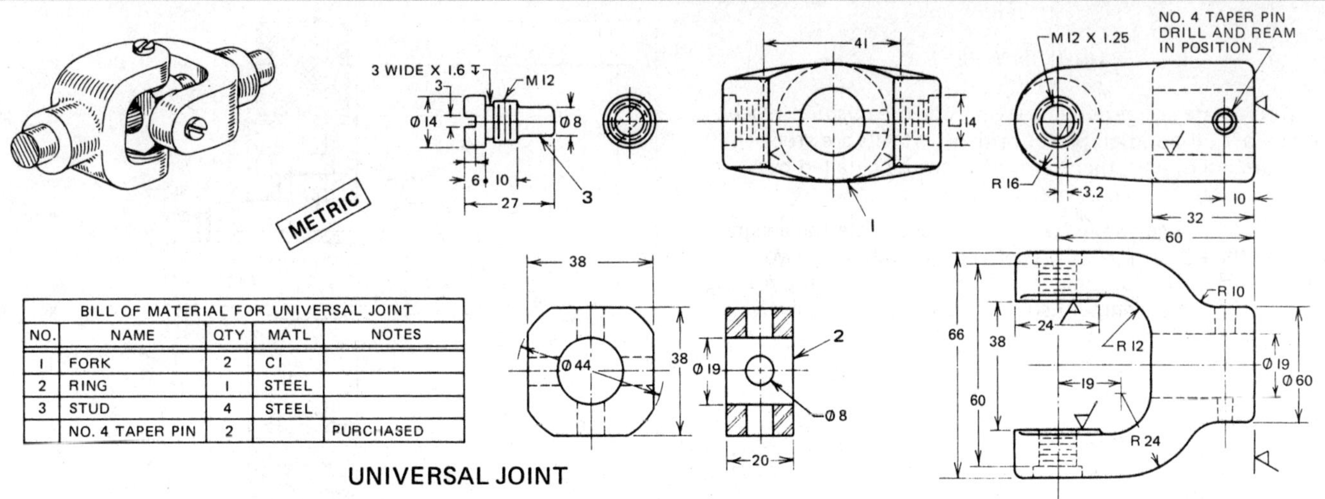

BILL OF MATERIAL FOR UNIVERSAL JOINT				
NO.	NAME	QTY	MATL	NOTES
1	FORK	2	CI	
2	RING	1	STEEL	
3	STUD	4	STEEL	
	NO. 4 TAPER PIN	2		PURCHASED

UNIVERSAL JOINT

Make a two-view assembly drawing of the *Universal Joint*. Show one view in half or full section. Scale: full size or as assigned by instructor.

WORKING DRAWINGS	DRAWN BY	DATE	DWG NO.

CUSHION WHEEL

Make a complete set of detail drawings for the *Cushion Wheel.* Include a bill of material. Scale: optional. Rivets are stock items (purchased) and, therefore, need not be detailed but must be listed in the bill of material.

Several drawing sheets may be needed to complete the assignment. Use the bottom of this sheet and the following two pages as needed. If additional sheets are needed, use the reverse side of any of the assignment sheets you have previously completed.

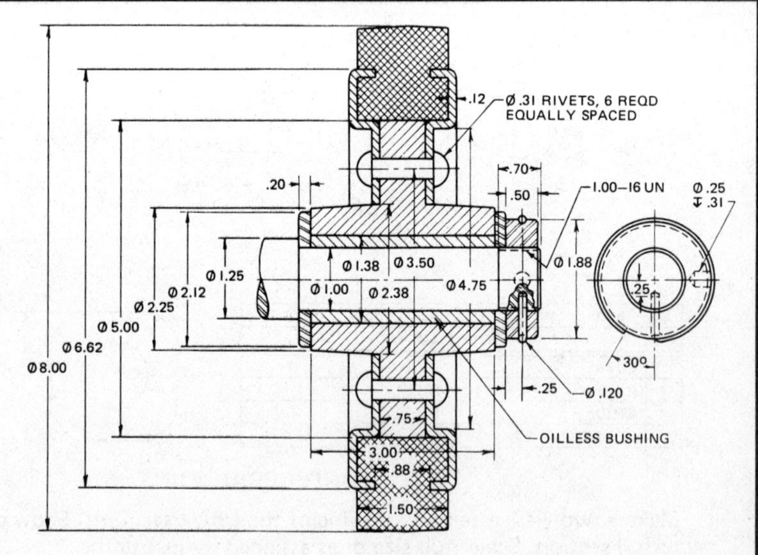

WORKING DRAWINGS	DRAWN BY	DATE	DWG NO.

WORKING DRAWINGS	DRAWN BY	DATE	DWG NO.

WORKING DRAWINGS	DRAWN BY	DATE	DWG NO.

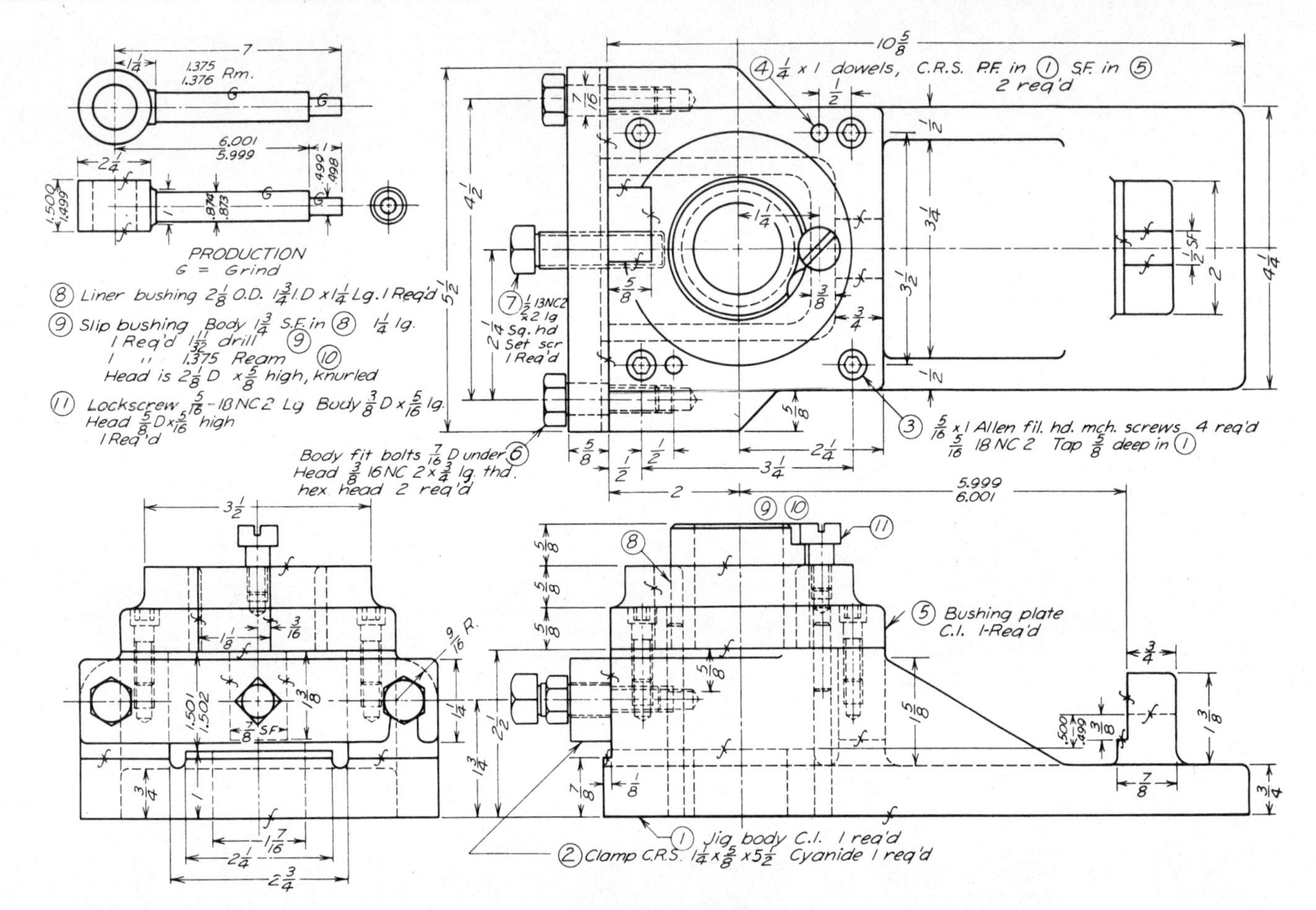

PRODUCTION JIG

A jig is a device used to hold a machine part (often called the work, production, or piece part) while it is being machined so that all finished parts will be alike within specified limits of accuracy. Study the drawing of the finished product in the upper left corner before beginning the assignments. Notice that all of the drawings above do not follow present ANSI standards. Be sure to make necessary changes to update each drawing as you complete the assignments. Several drawing sheets will be needed to complete the assignments. Use the bottom of this sheet and the following three pages. If additional sheets are needed, use the reverse side of any of the assignment sheets you have previously completed.

ASSIGNMENT 1: Make a detail drawing of the jig body. Scale: optional.

ASSIGNMENT 2: Make a complete set of detail working drawings for the jig. Include a bill of material.

ASSIGNMENT 3: Make a complete three-view assembly drawing of the jig. The production (work, piece part) may be shown in place on the jig using phantom lines. Give only such dimensions as are needed for assembling and using the jig.

WORKING DRAWINGS	DRAWN BY	DATE	DWG NO.

Production Jig

WORKING DRAWINGS	DRAWN BY	DATE	DWG NO.

Production Jig

WORKING DRAWINGS	DRAWN BY	DATE	DWG NO.

Production Jig

WORKING DRAWINGS	DRAWN BY	DATE	DWG NO.

Use instruments to make isometric drawings of the objects below. A beginning point (corner) is given for each.
Use dividers to take sizes from the multiview drawings and transfer them double size to the isometric drawing.

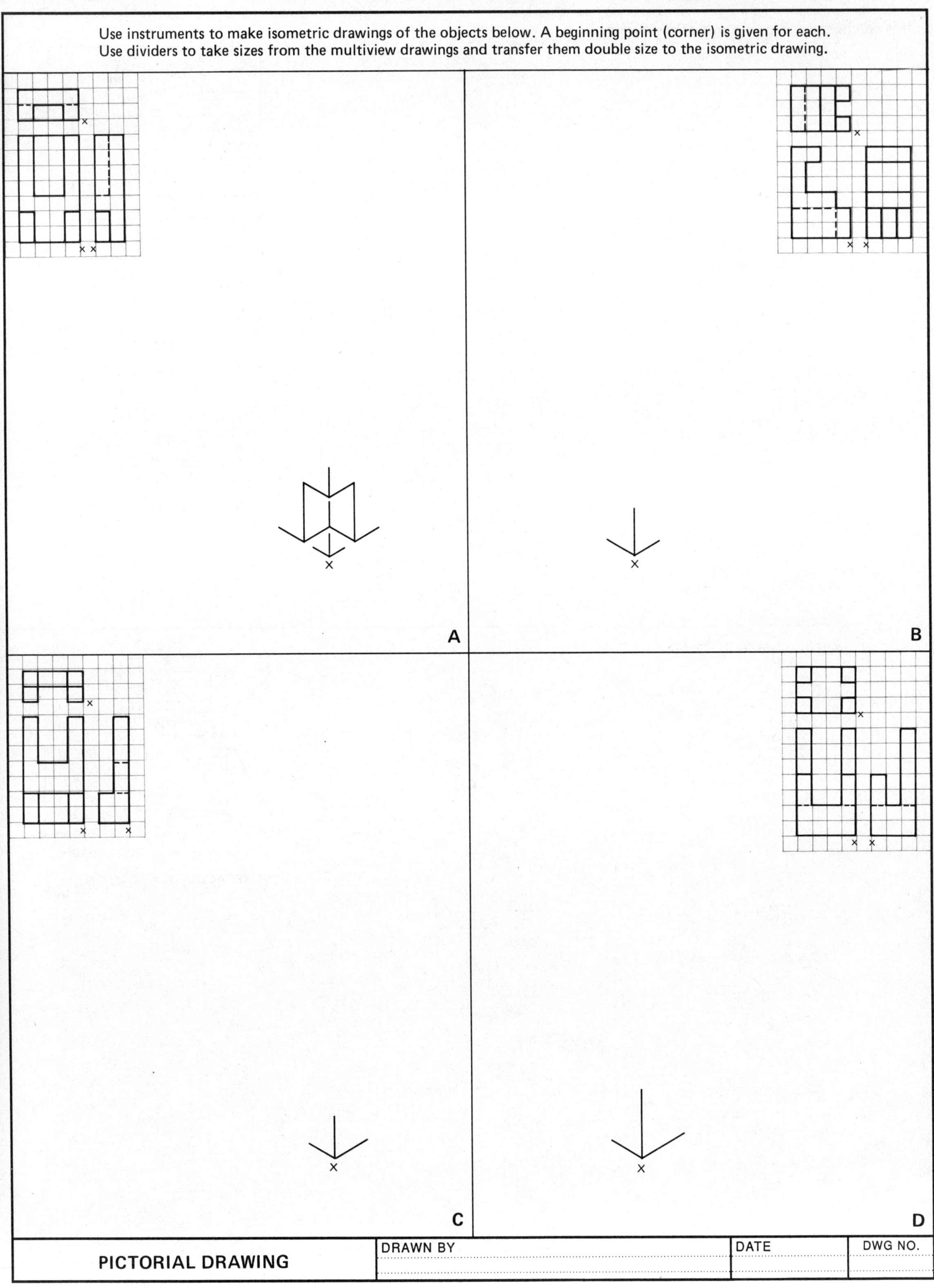

PICTORIAL DRAWING	DRAWN BY	DATE	DWG NO.

Make an isometric drawing of the *Saw Bracket*.

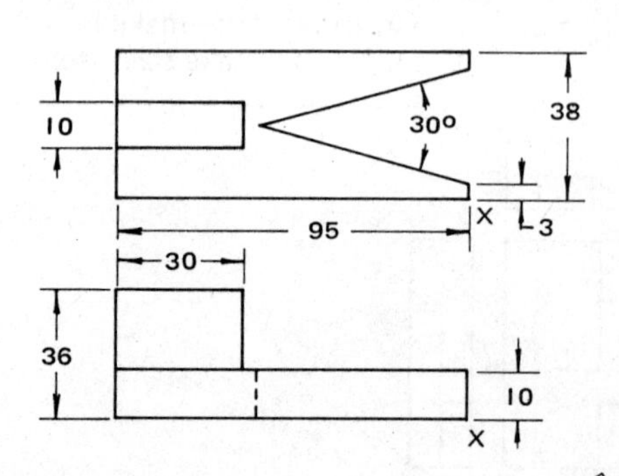

Make an isometric drawing of the *Brace*.

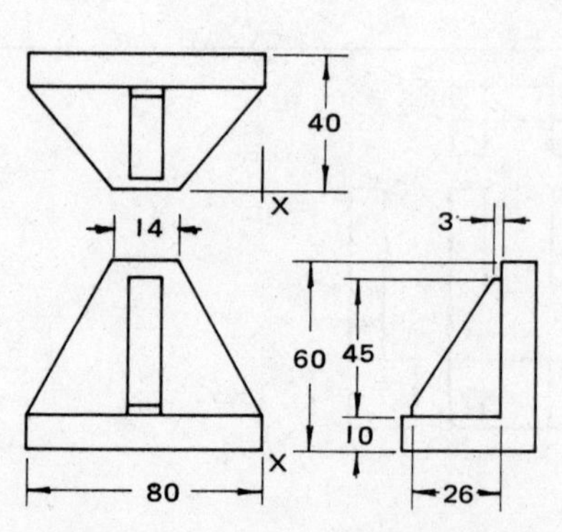

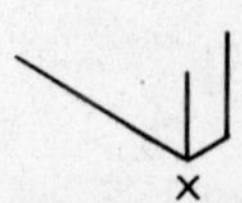

PICTORIAL DRAWING	DRAWN BY	DATE	DWG NO.

Make an isometric drawing of the *Cross Slide.*

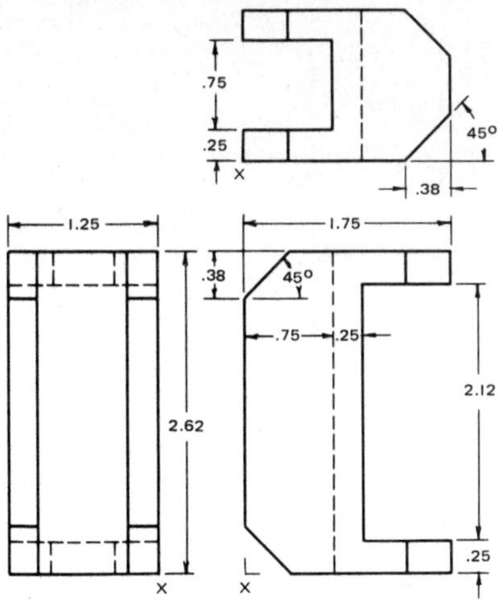

CROSS SLIDE

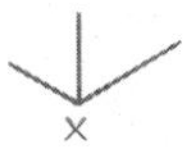

A

Make a half-size isometric drawing of the *Ratchet.*

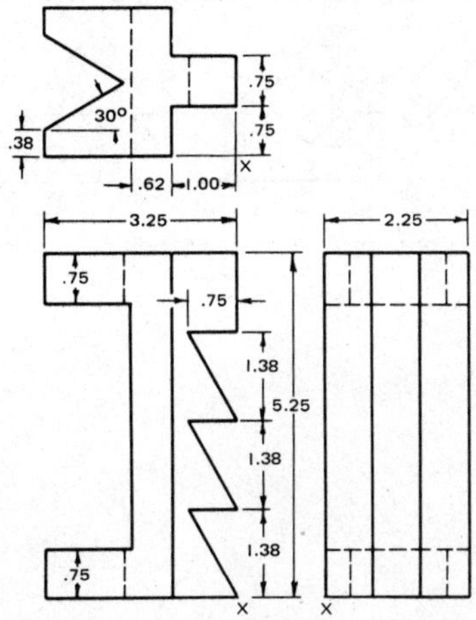

RATCHET

B

PICTORIAL DRAWING	DRAWN BY	DATE	DWG NO.

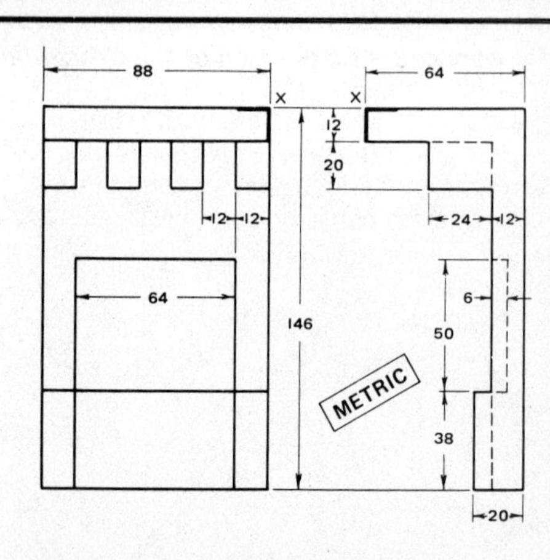

Make an isometric drawing of the *Tablet*. Use reversed axes. Refer to the layout below.

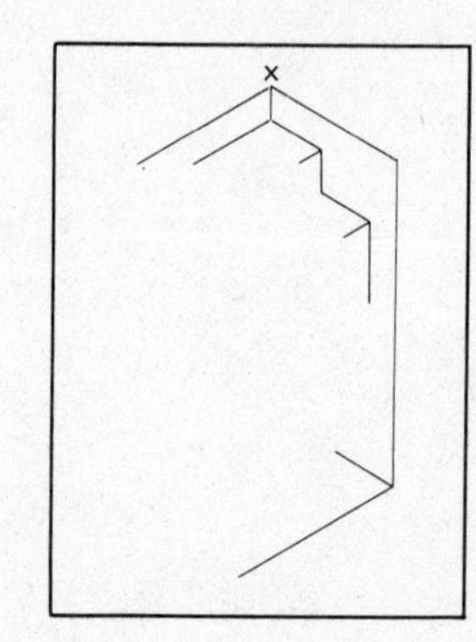

A

Make an isometric drawing of the *Vise Jaw*.

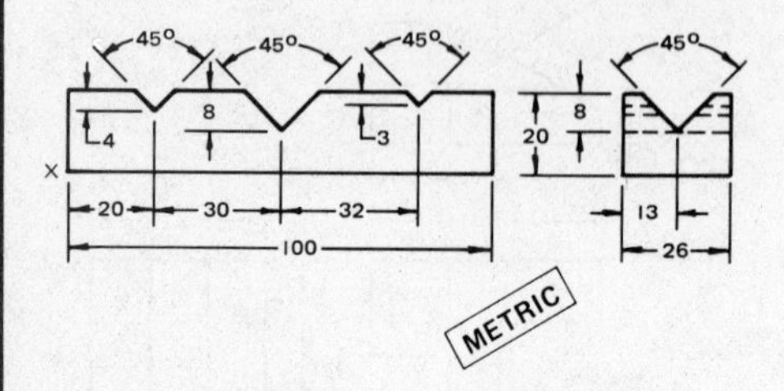

B

PICTORIAL DRAWING	DRAWN BY	DATE	DWG NO.

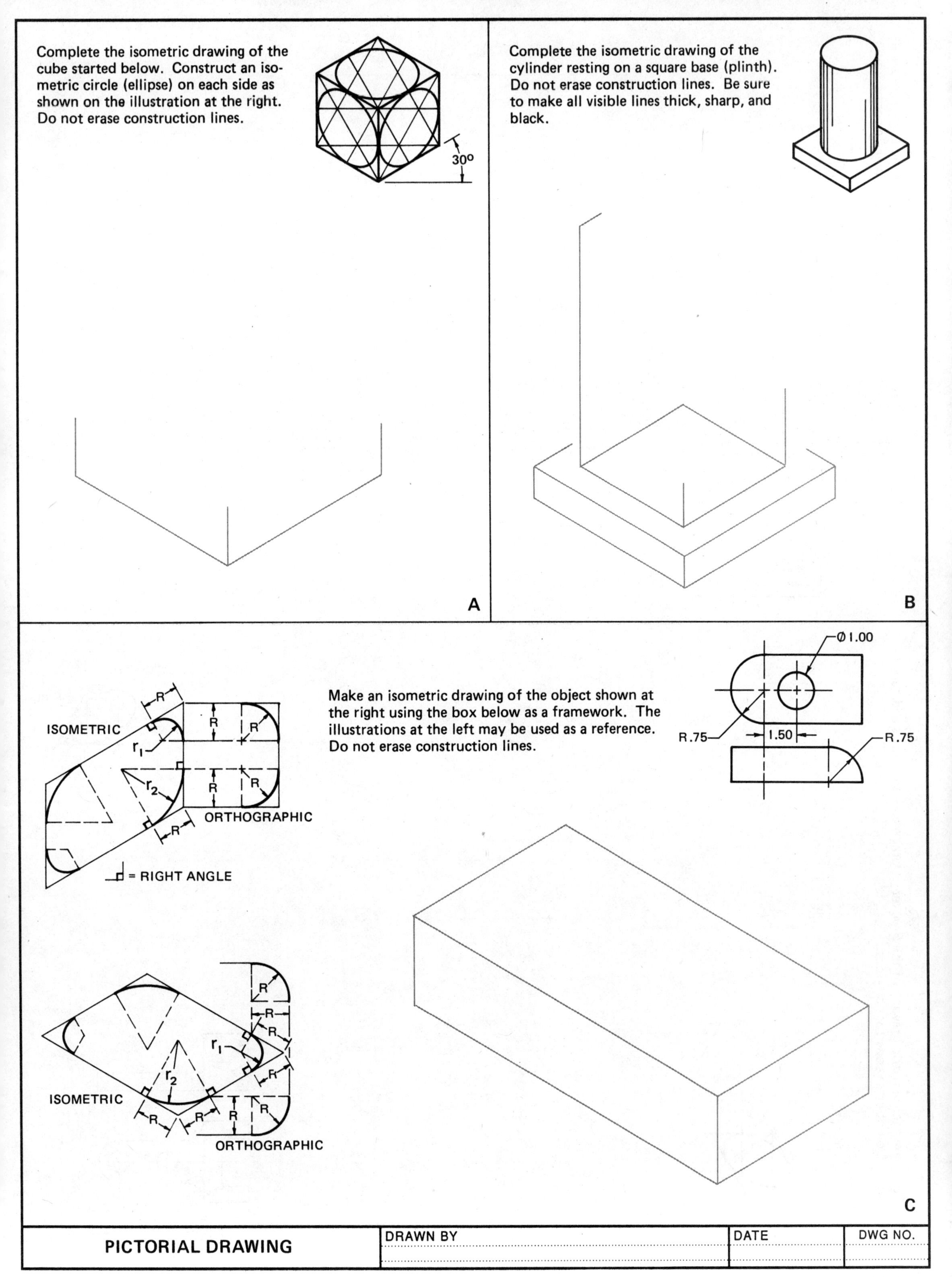

Complete the isometric drawing of the cube started below. Construct an isometric circle (ellipse) on each side as shown on the illustration at the right. Do not erase construction lines.
30°
A
Complete the isometric drawing of the cylinder resting on a square base (plinth). Do not erase construction lines. Be sure to make all visible lines thick, sharp, and black.
B
ISOMETRIC
R
r1
r2
ORTHOGRAPHIC
= RIGHT ANGLE
Make an isometric drawing of the object shown at the right using the box below as a framework. The illustrations at the left may be used as a reference. Do not erase construction lines.
Ø1.00
R.75
1.50
R.75
ISOMETRIC
ORTHOGRAPHIC
C
PICTORIAL DRAWING
DRAWN BY
DATE
DWG NO.

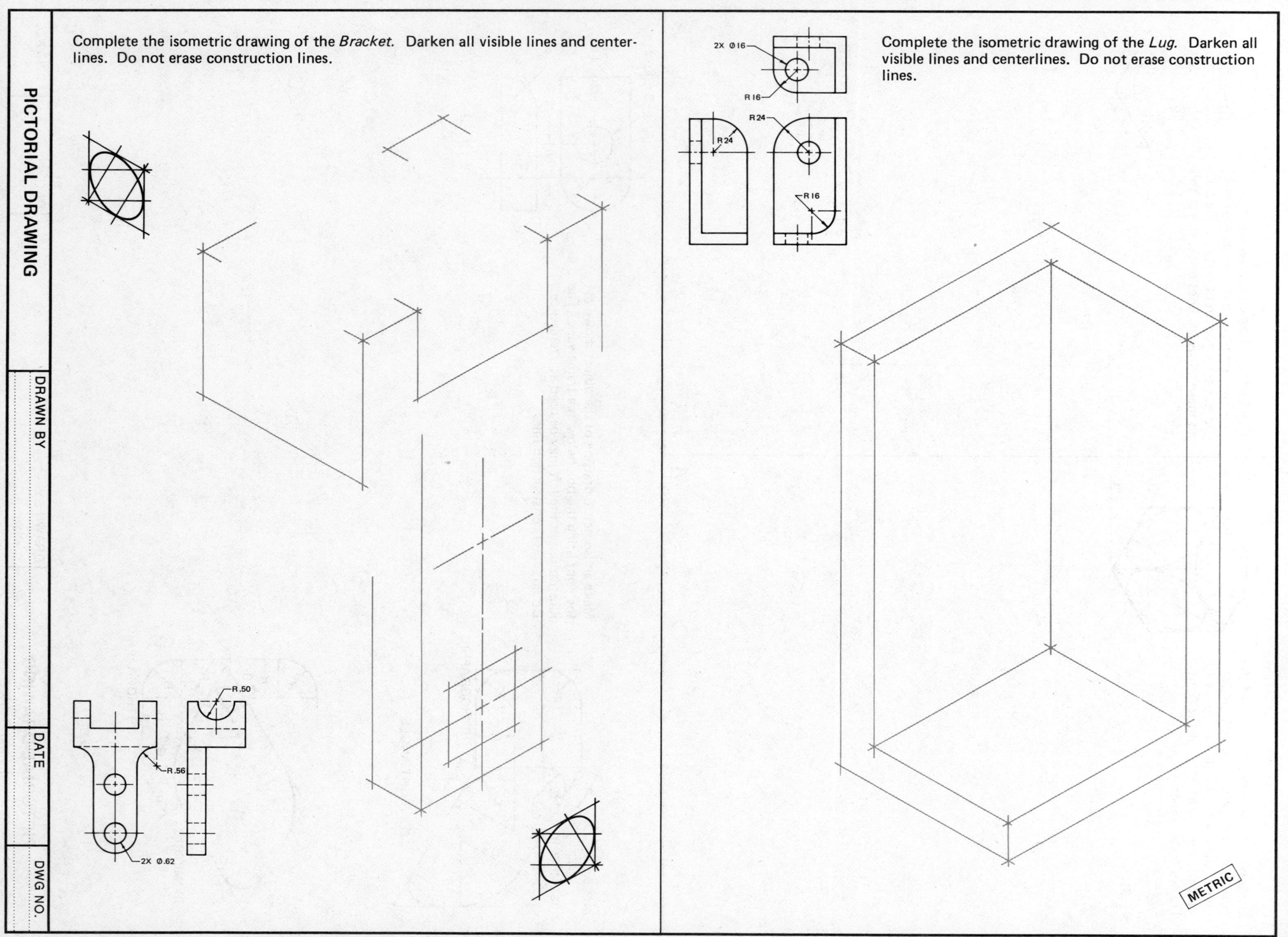

PICTORIAL DRAWING

DRAWN BY	DATE	DWG NO.

Transfer points from the grid at the right to the grid below and complete the isometric drawing of the *Brace.* Use an irregular curve to connect points.

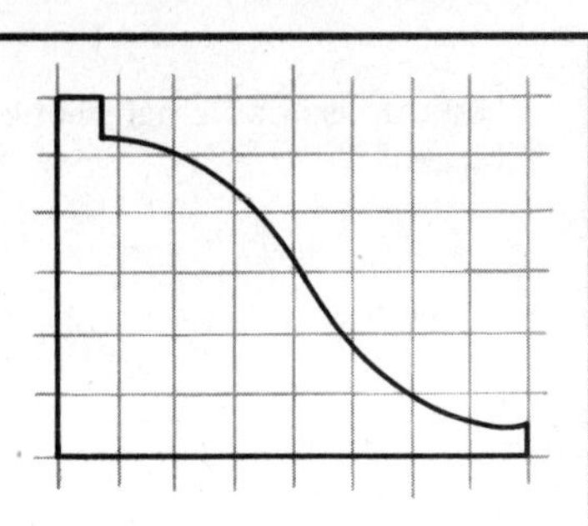

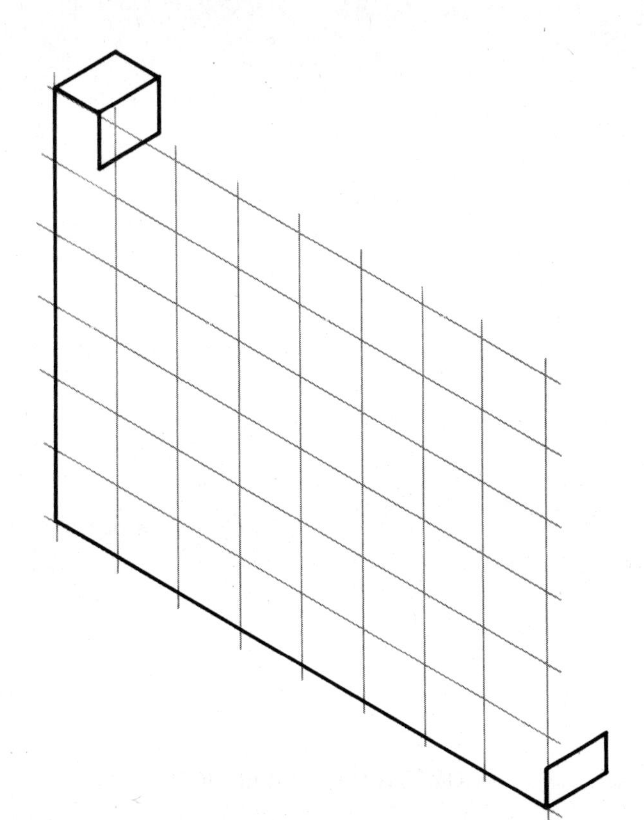

A

Develop an isometric drawing of the *Lever* using a grid to transfer points. Begin the grid at point X and double the size of the squares. Use an irregular curve to connect points and either an ellipse template or the four-center method to draw the circular features (ellipses).

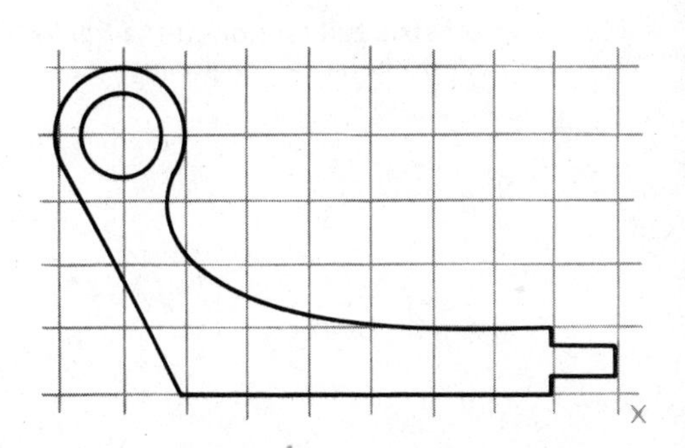

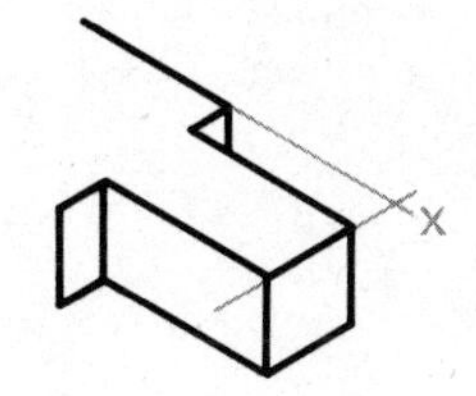

B

PICTORIAL DRAWING	DRAWN BY	DATE	DWG NO.

Make an isometric half section and an isometric full section of the *Rod Stand.*
Scale 1 : 1

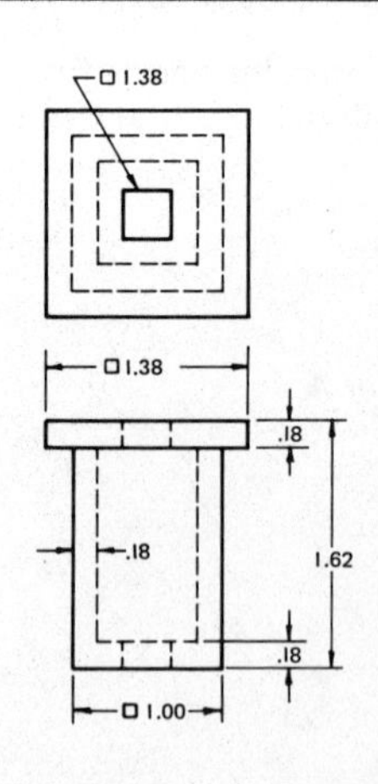

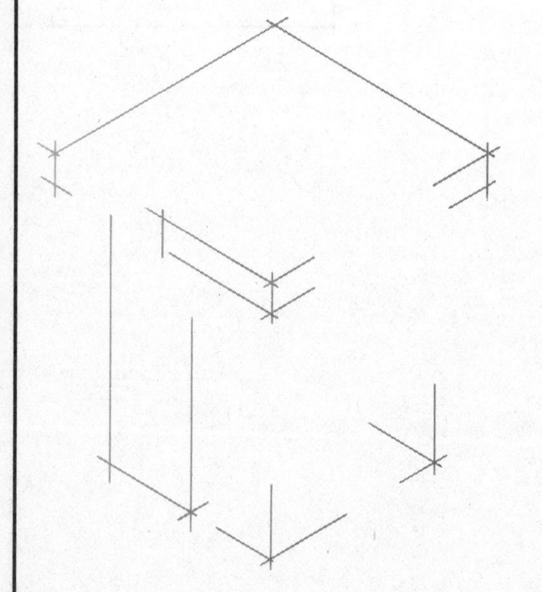

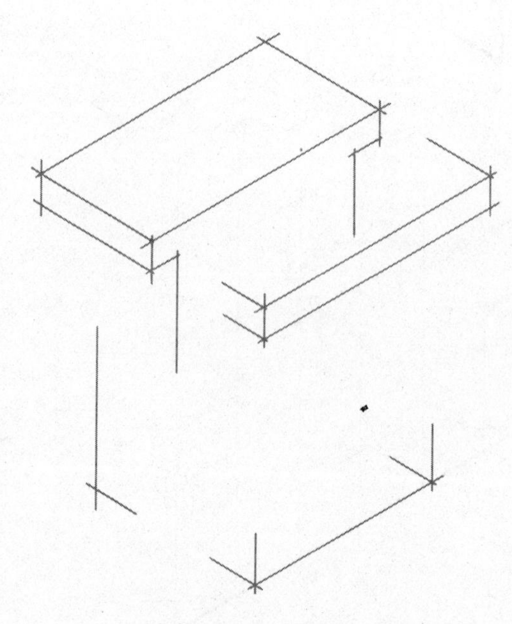

ISOMETRIC HALF SECTION

ISOMETRIC FULL SECTION

A

Make an isometric full section of the *Baffle.* Scale 1 : 1. Study the front and top views carefully before beginning.

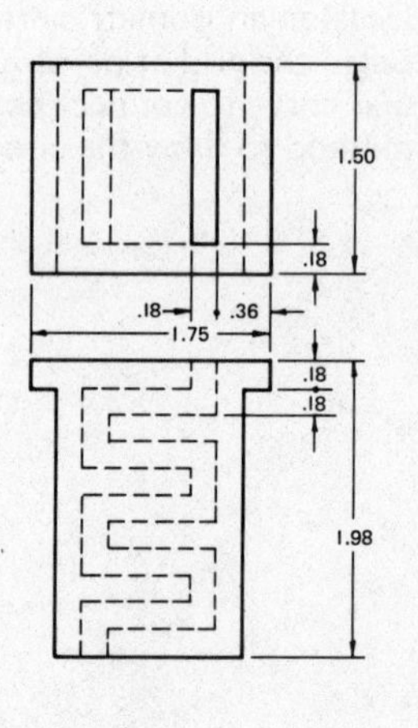

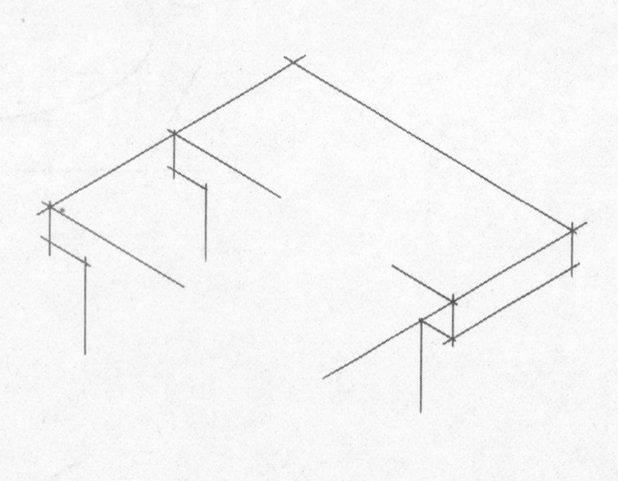

B

PICTORIAL DRAWING	DRAWN BY	DATE	DWG NO.

Make an isometric half section and an isometric full section of the *Leveling Cone.* Scale 1 : 1. Use general-purpose section lining. Ellipses may be drawn using the four-center method or an ellipse template.

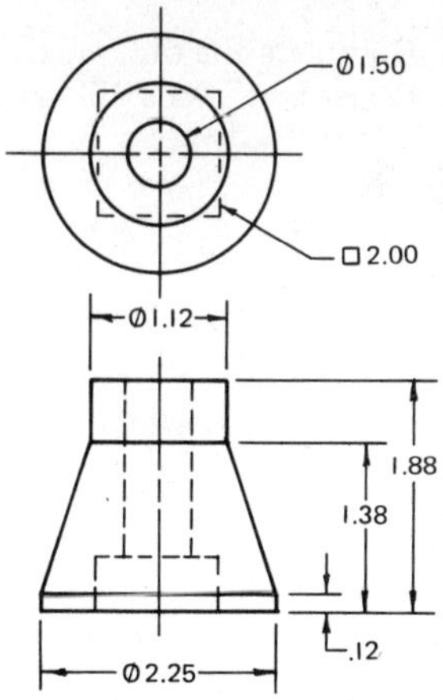

LEVELING CONE

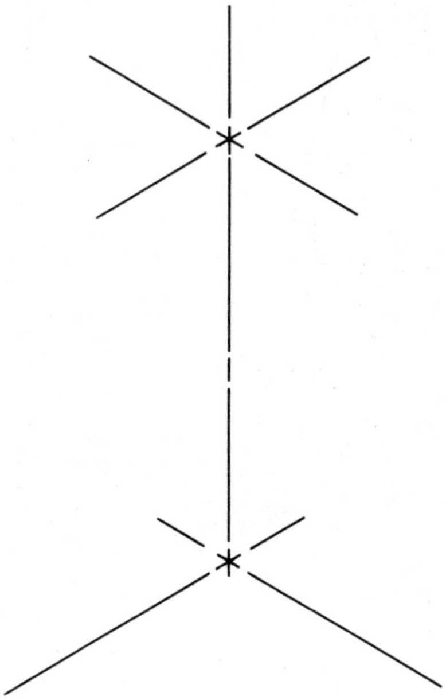

ISOMETRIC HALF SECTION

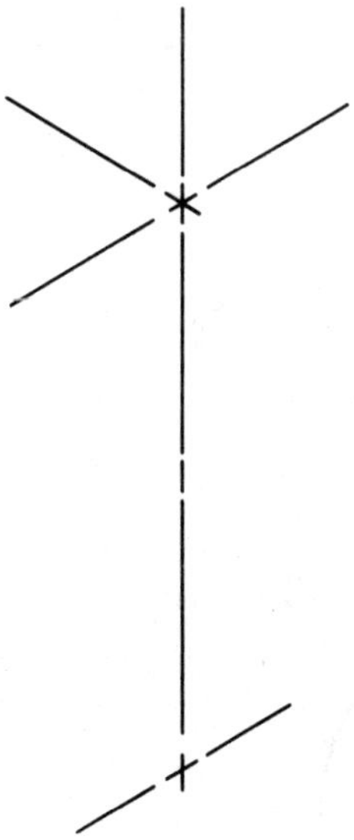

ISOMETRIC FULL SECTION

A

Make an isometric half section and an isometric full section of the *Centering Cone.* Scale 1 : 1. Use general-purpose section lining. Ellipses may be drawn using the four-center method or an ellipse template.

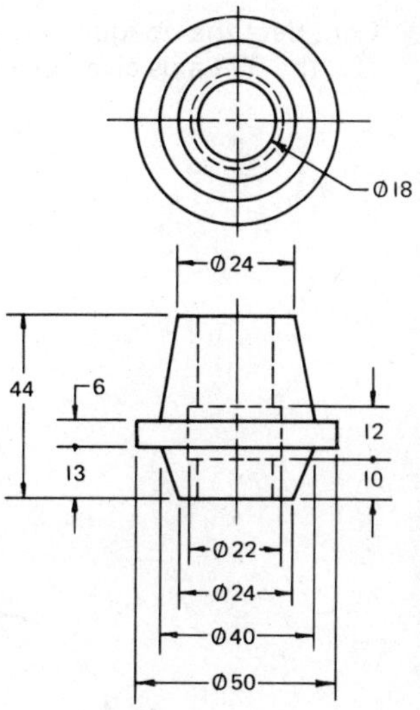

CENTERING CONE

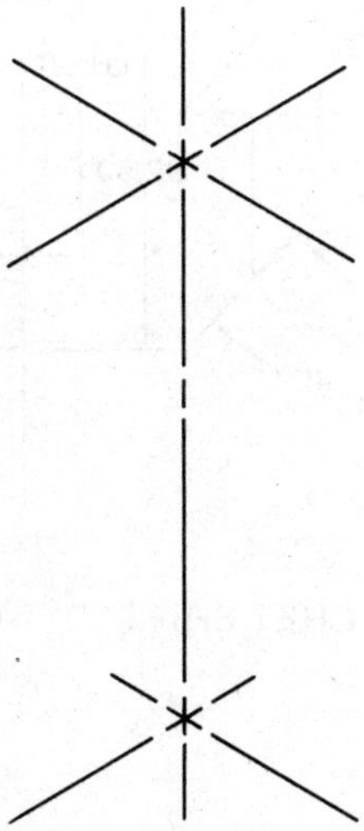

ISOMETRIC HALF SECTION

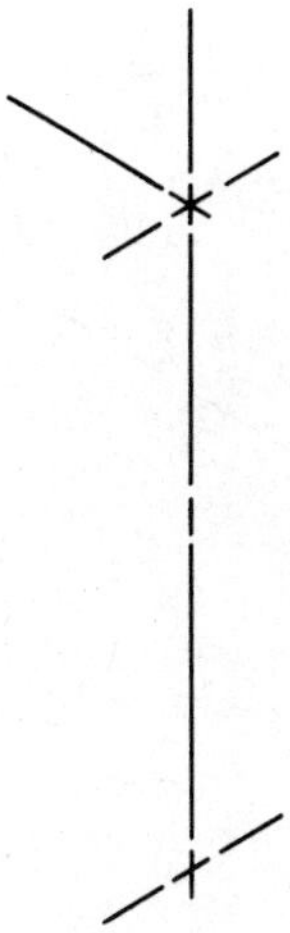

ISOMETRIC FULL SECTION

METRIC

B

PICTORIAL DRAWING	DRAWN BY	DATE	DWG NO.

Complete the oblique cavalier drawing of the *Crank*.
Scale 1 : 1. Axis 45° to the right.

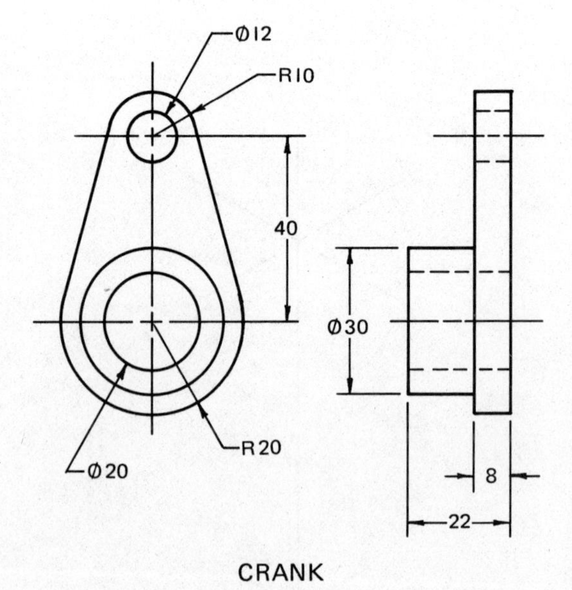

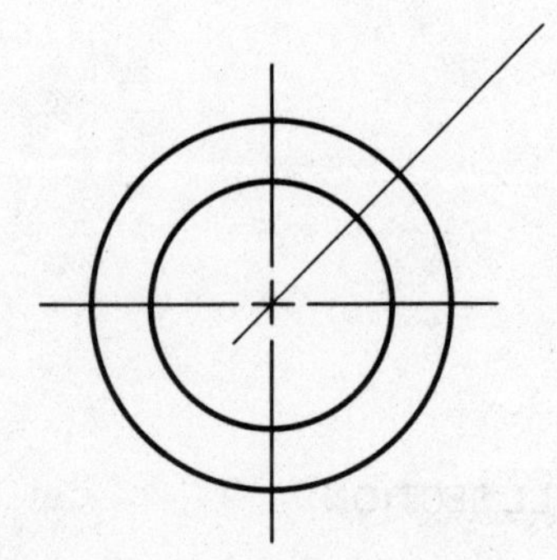

METRIC

A

Complete the oblique cavalier drawing of the *Ratchet Shifter*.
Use the 30° axis given below. Scale 1 : 1.

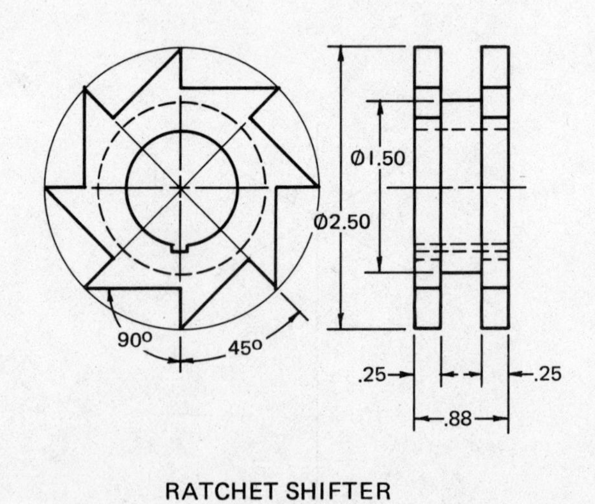

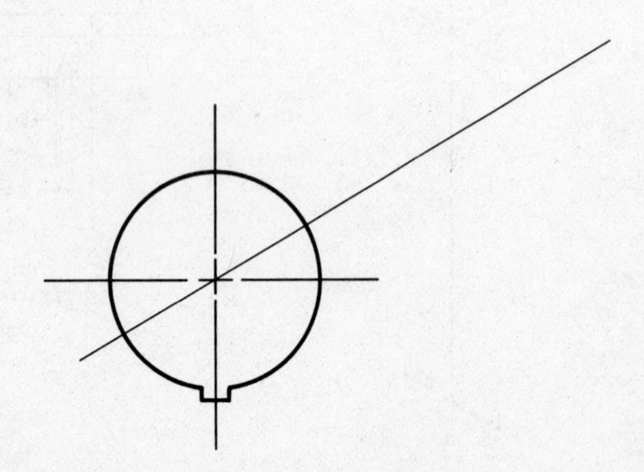

B

PICTORIAL DRAWING	DRAWN BY	DATE	DWG NO.

Make an oblique drawing of the *Centering Cone.* Use the 30° axis given below. Scale 1 : 1.

Make an oblique half section of the *Centering Cone.* Use the 45° axis given below. Scale 1 : 1.

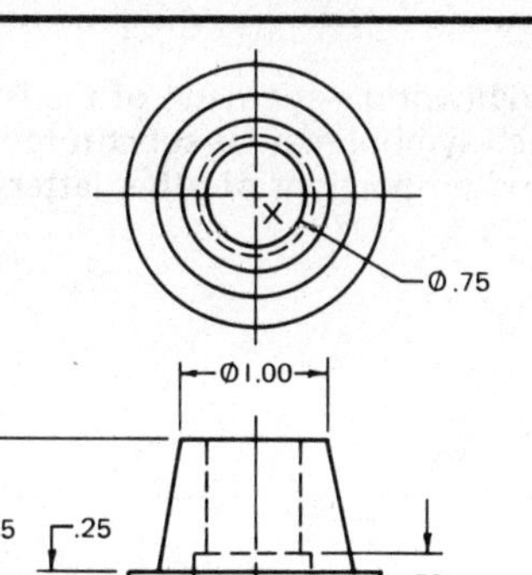

CENTERING CONE

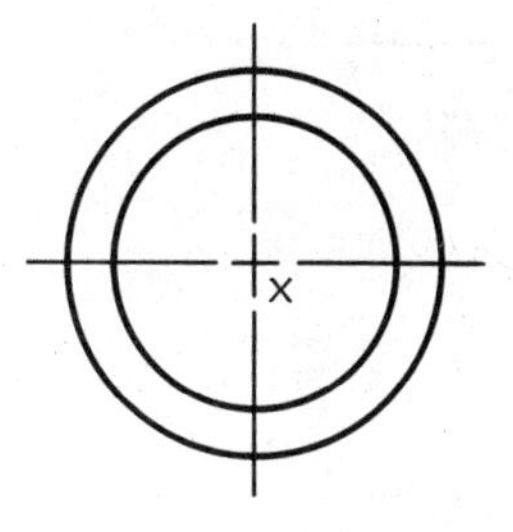

OBLIQUE DRAWING

A

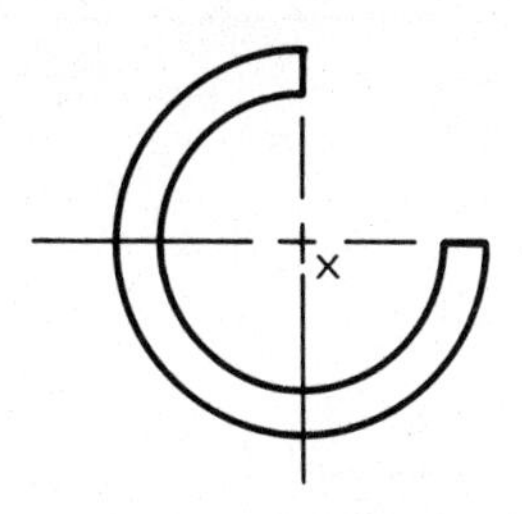

OBLIQUE HALF SECTION

B

Complete the oblique drawing below using information obtained from the orthographic drawing at the right. Centerlines for circles and arcs are shown. Extreme accuracy is required in drawing the small holes. Review the unit on oblique drawing in your textbook before beginning this assignment.

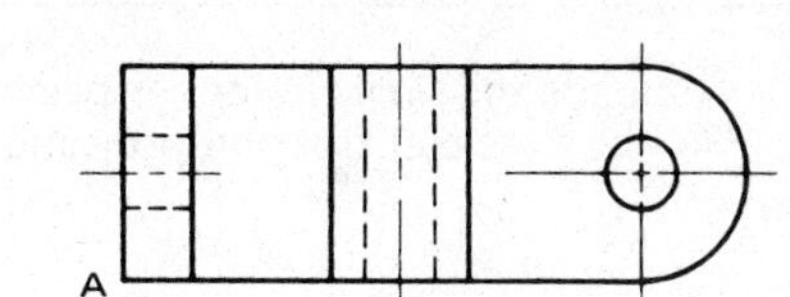

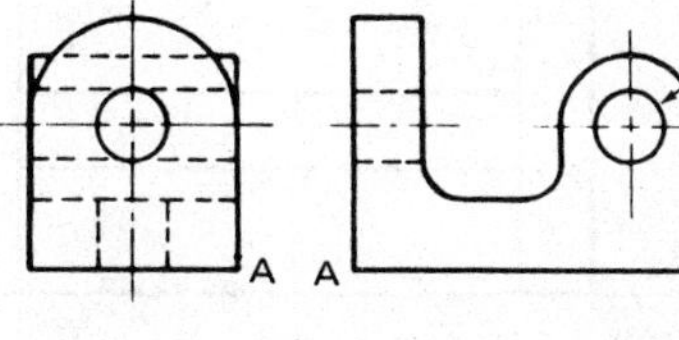

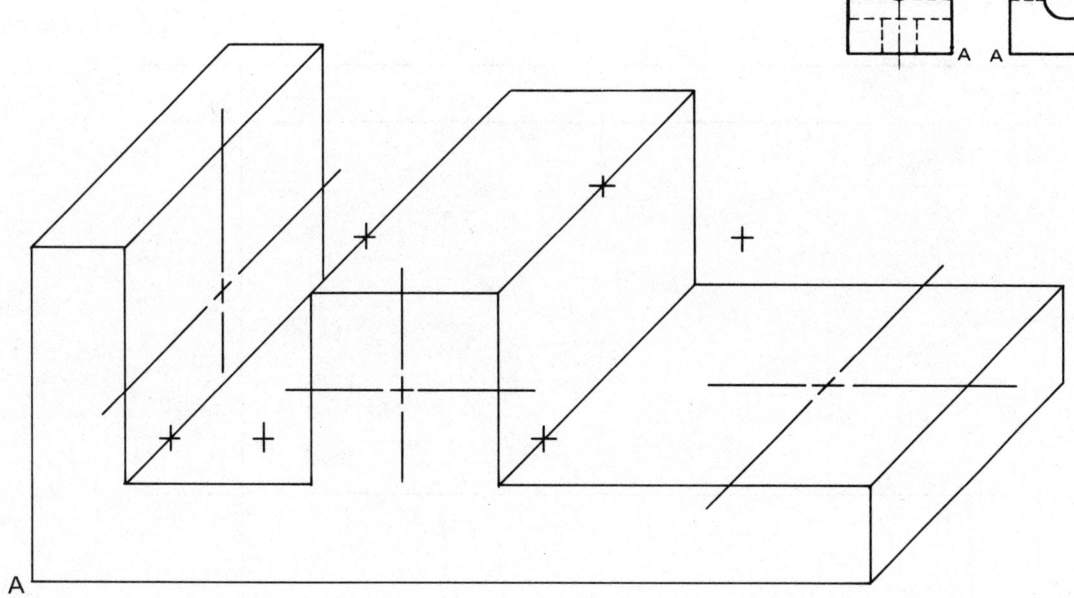

C

PICTORIAL DRAWING	DRAWN BY	DATE	DWG NO.

Complete the single-point perspective drawing (worm's-eye view) of the letter N. Your own initial(s), school logo, or other such symbol may be substituted. See Chapter 2 in your textbook for the design and proportions of other letters. Do not erase construction/projection lines. Scale 1 : 1.

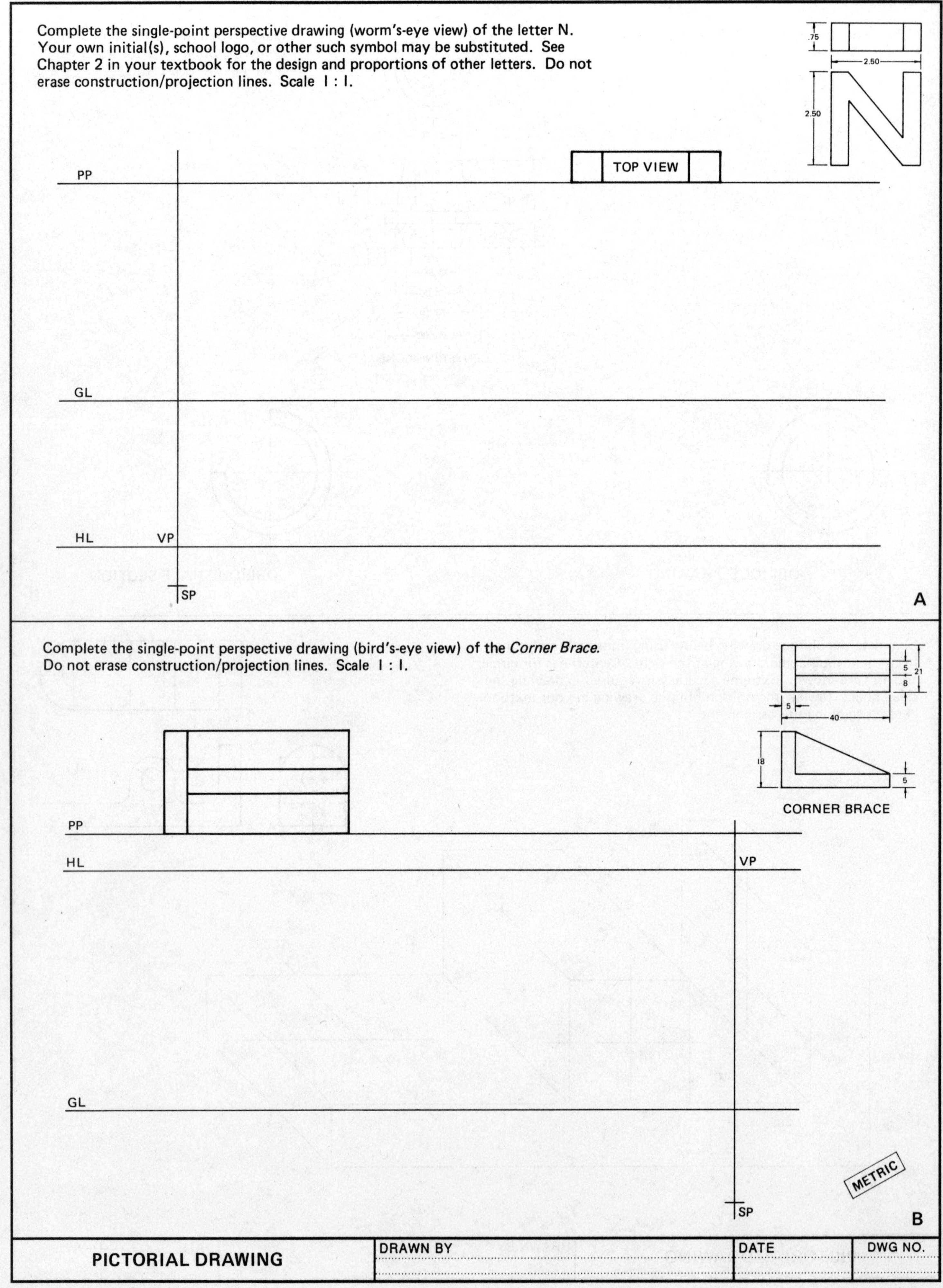

Complete the single-point perspective drawing (bird's-eye view) of the *Corner Brace*. Do not erase construction/projection lines. Scale 1 : 1.

PICTORIAL DRAWING	DRAWN BY	DATE	DWG NO.

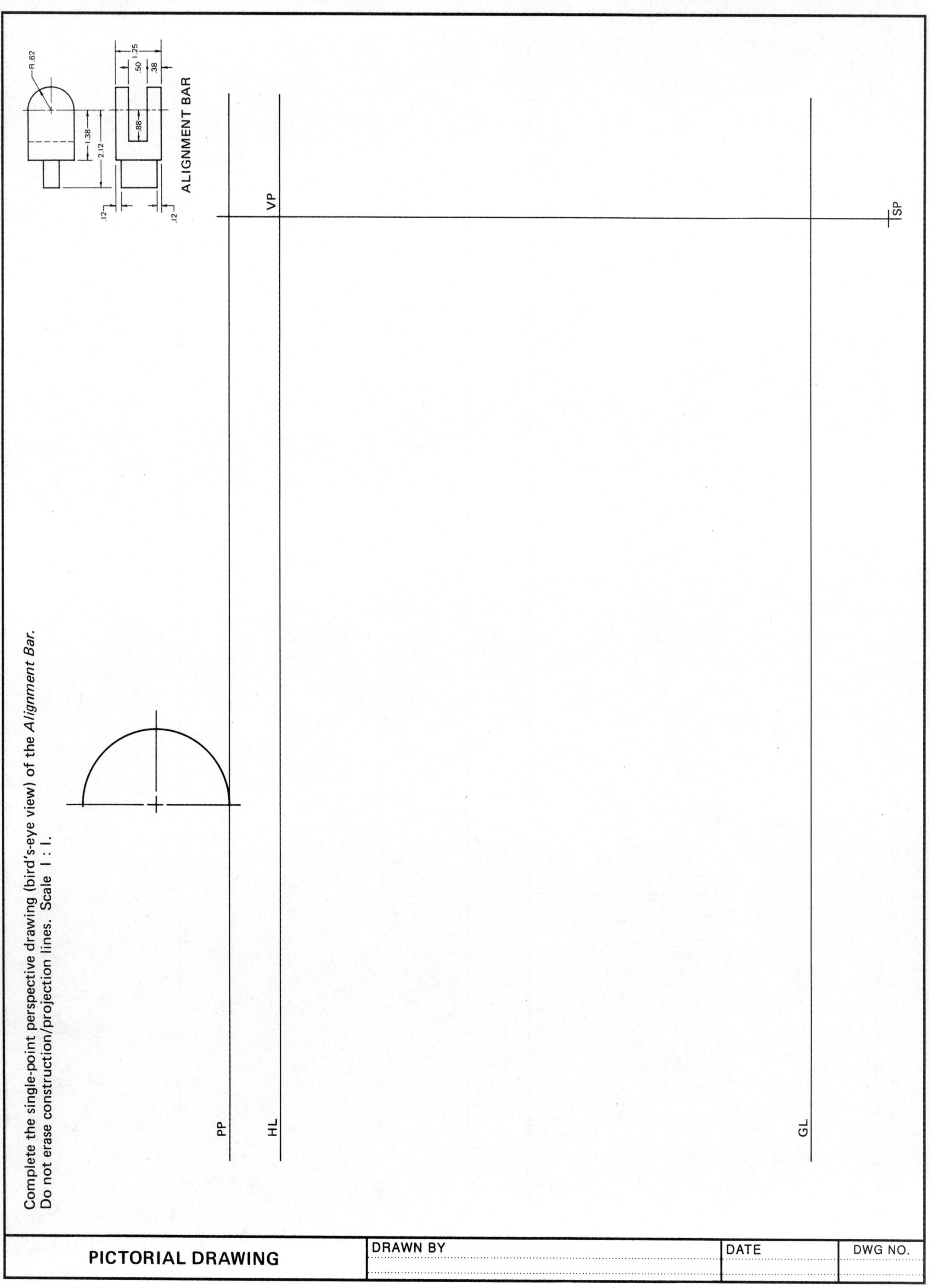

Complete the single-point perspective drawing (bird's-eye view) of the *Alignment Bar*.
Do not erase construction/projection lines. Scale 1 : 1.
ALIGNMENT BAR
R.62
1.25
.50
.38
1.38
2.12
.88
.12
.12
VP
SP
PP
HL
GL
PICTORIAL DRAWING
DRAWN BY
DATE
DWG NO.

Complete the two-point perspective drawing (bird's-eye view) of the *Dovetail Slide.*
Do not erase construction/projection lines. Scale 1 : 1.

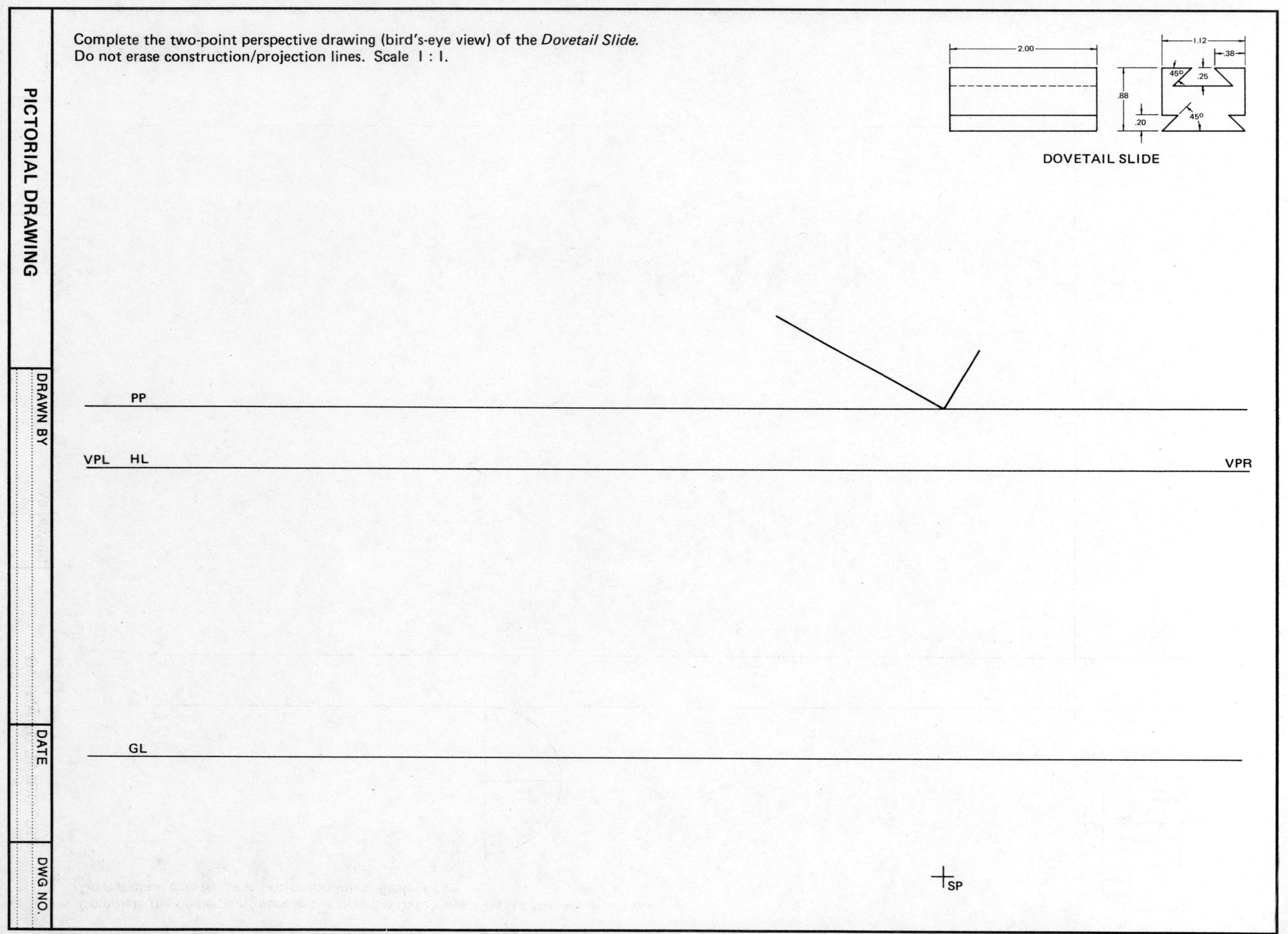

PICTORIAL DRAWING	DRAWN BY	DATE	DWG NO.

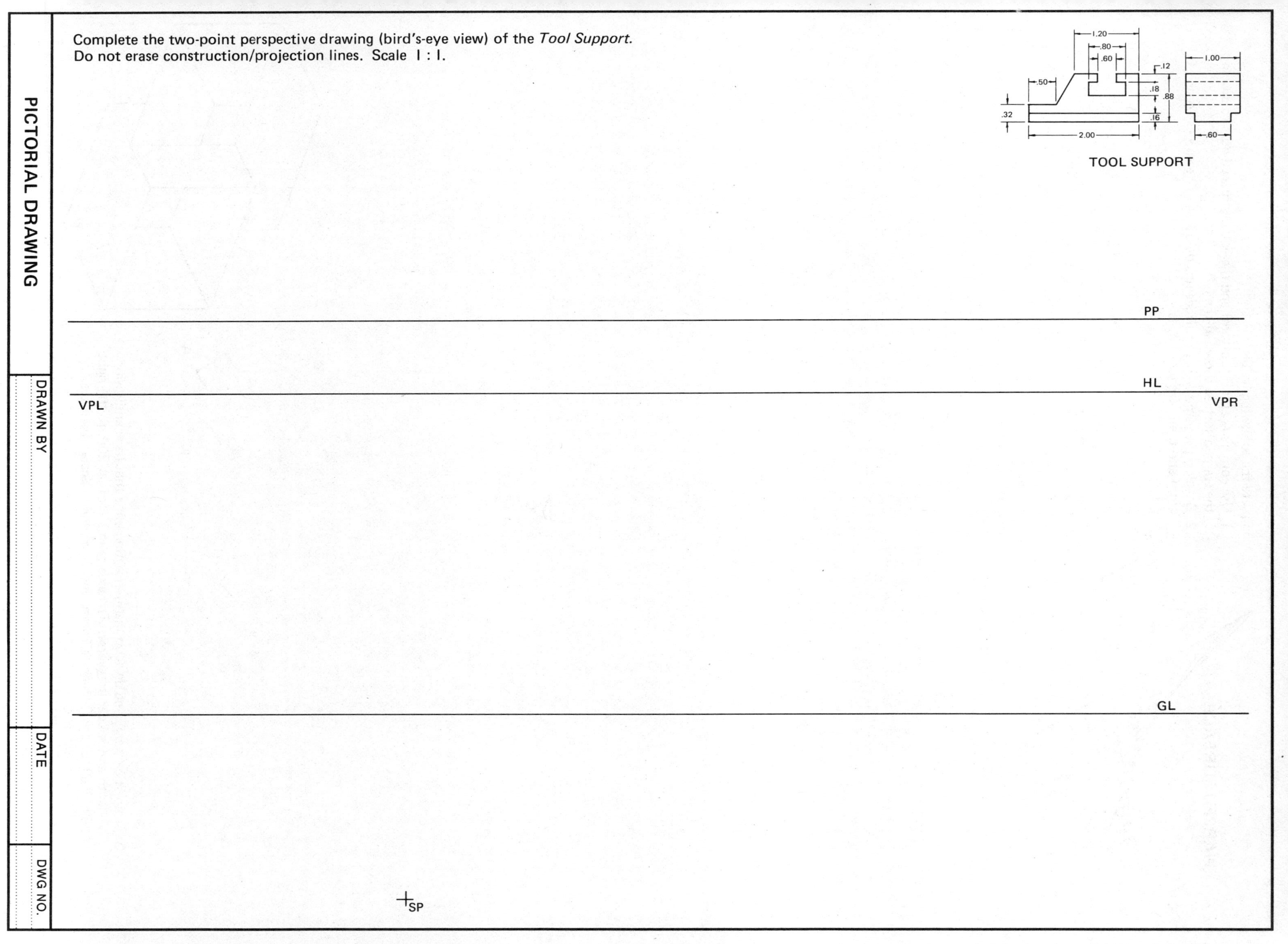
Complete the two-point perspective drawing (bird's-eye view) of the *Tool Support*.
Do not erase construction/projection lines. Scale 1 : 1.

PICTORIAL DRAWING	DRAWN BY	DATE	DWG NO.

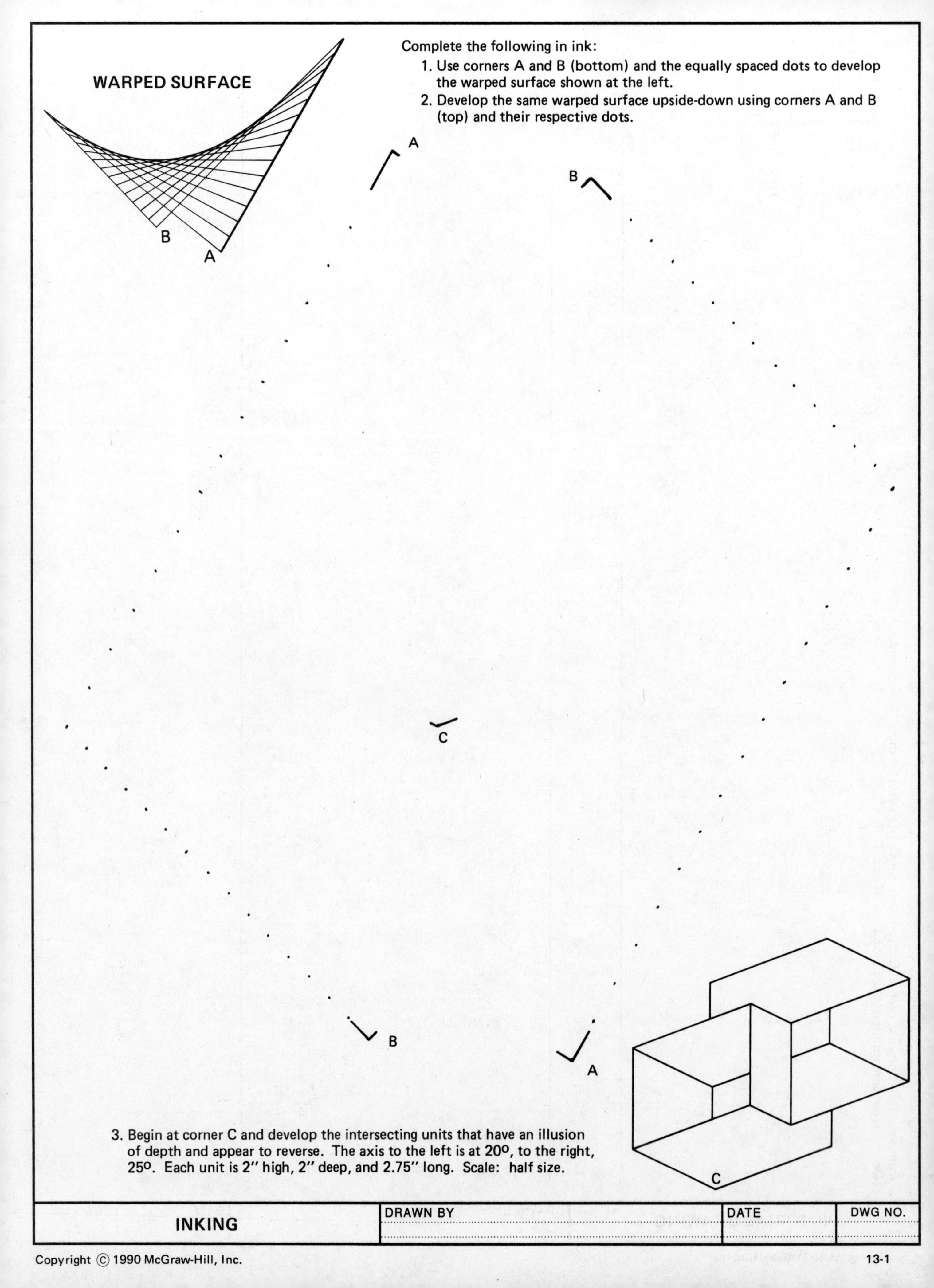

Complete the following in ink:

1. Use corners A and B (bottom) and the equally spaced dots to develop the warped surface shown at the left.
2. Develop the same warped surface upside-down using corners A and B (top) and their respective dots.

3. Begin at corner C and develop the intersecting units that have an illusion of depth and appear to reverse. The axis to the left is at 20°, to the right, 25°. Each unit is 2" high, 2" deep, and 2.75" long. Scale: half size.

INKING	DRAWN BY	DATE	DWG NO.

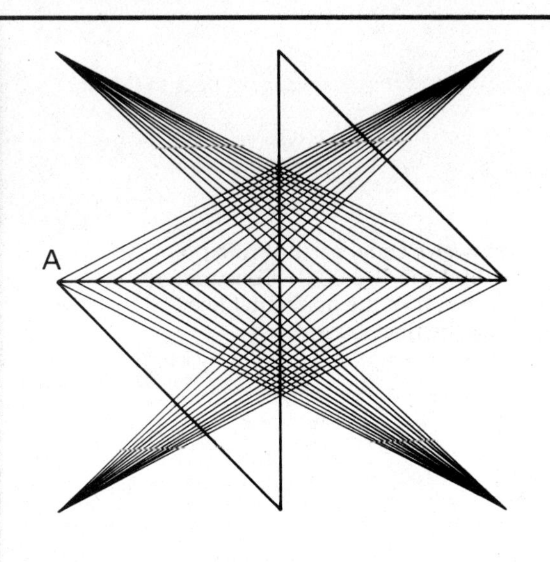

Using light pencil construction lines, lay off a 6″ square and divide it equally into four 3″ squares. On the horizontal midline, mark off 24 one-quarter inch units. Proceed to ink in the lines to construct the given design.

A_

INKING	DRAWN BY	DATE	DWG NO.

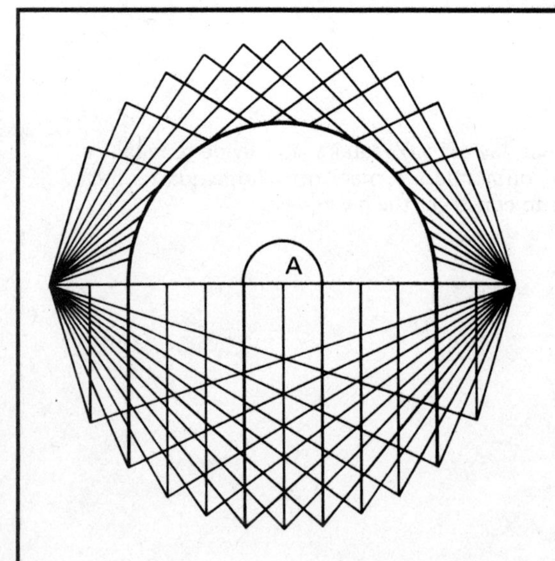

Divide the 6″ horizontal line into twelve equal parts. From point A, construct the two arcs shown (R .50″ and R 2.00″). Construct a circle around point A touching each end of the horizontal line. Proceed to develop the design. Ink the final drawing.

A

INKING	DRAWN BY	DATE	DWG NO.

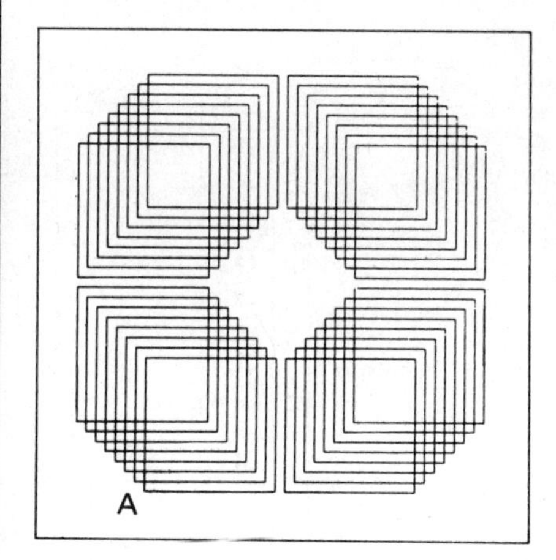

Beginning at point A, develop 36 two-inch squares, one-eighth inch apart. Ink the final drawing. The optional design may be done on another drawing sheet if assigned.

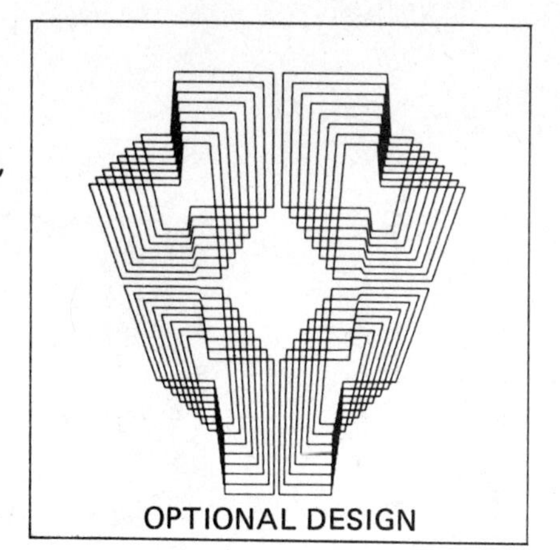

OPTIONAL DESIGN

A

INKING	DRAWN BY	DATE	DWG NO.

Create a CAD drawing of the *Dovetail Form Tool.* Select an appropriate scale for the contoured cutting-edge detail.

Plot your finished drawing on the reverse side of an earlier completed drawing or on a separate sheet.

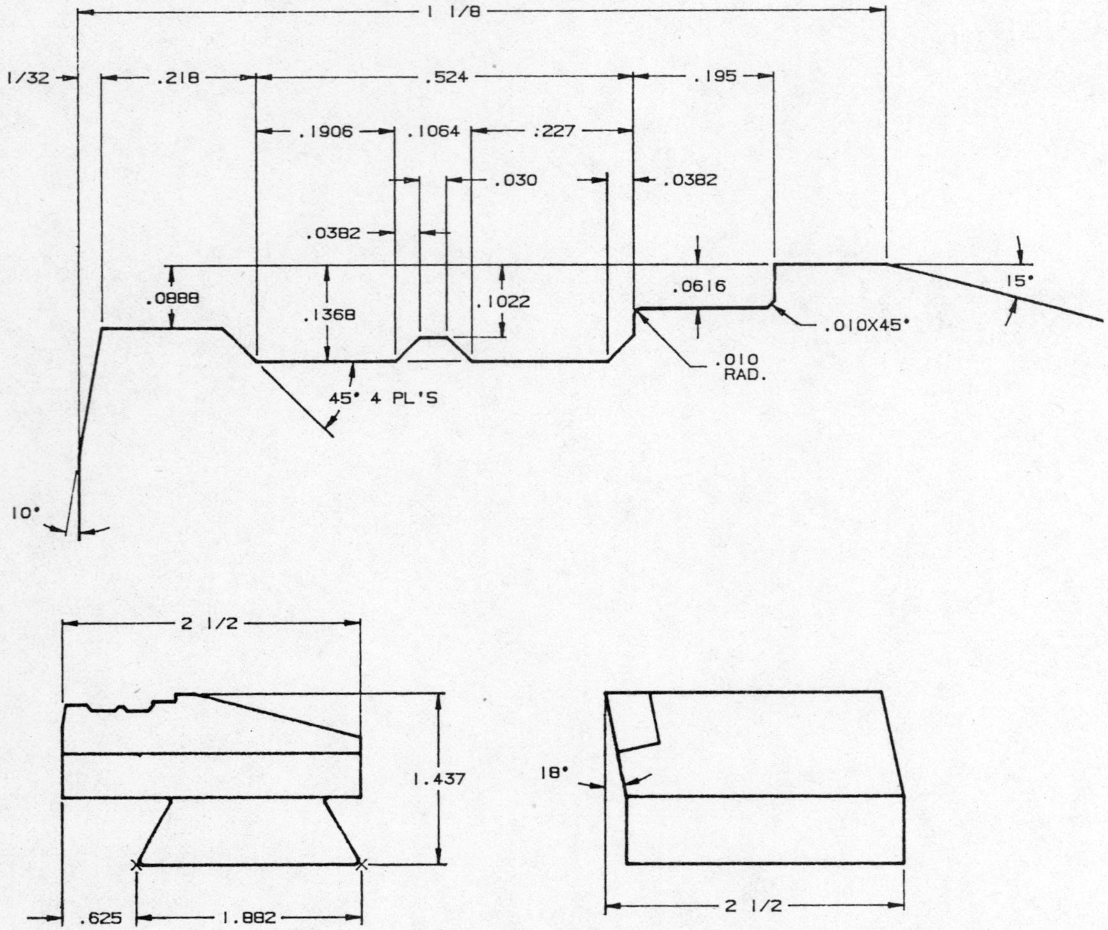

C-6 CARBIDE, T.S. HD'N. & GR'D.

DET.	REQ'D.	FINISH SIZE	MAT'L.

THE PYLE-NATIONAL COMPANY
CHICAGO, ILLINOIS 60651

T.C.T. DOVETAIL FORM TOOL

UNLESS OTHERWISE SHOWN TOLERANCES

FRACTIONAL ±1/64
DECIMAL ±.001
ANGULAR ±0°15'

DR.	DATE	DWG. NO.
CHKD.	APP.	

REV.	DATE	BY	E.C.O.	M.C.O.	CHANGES

COMPUTER-AIDED DRAFTING AND DESIGN

DRAWN BY

DATE

DWG NO.

Create a CAD drawing of the *Dovetail Form Tool*. Plot your finished drawing on the reverse side of this sheet. Scale the drawing to fit the space available. Add notes and other details as appropriate.

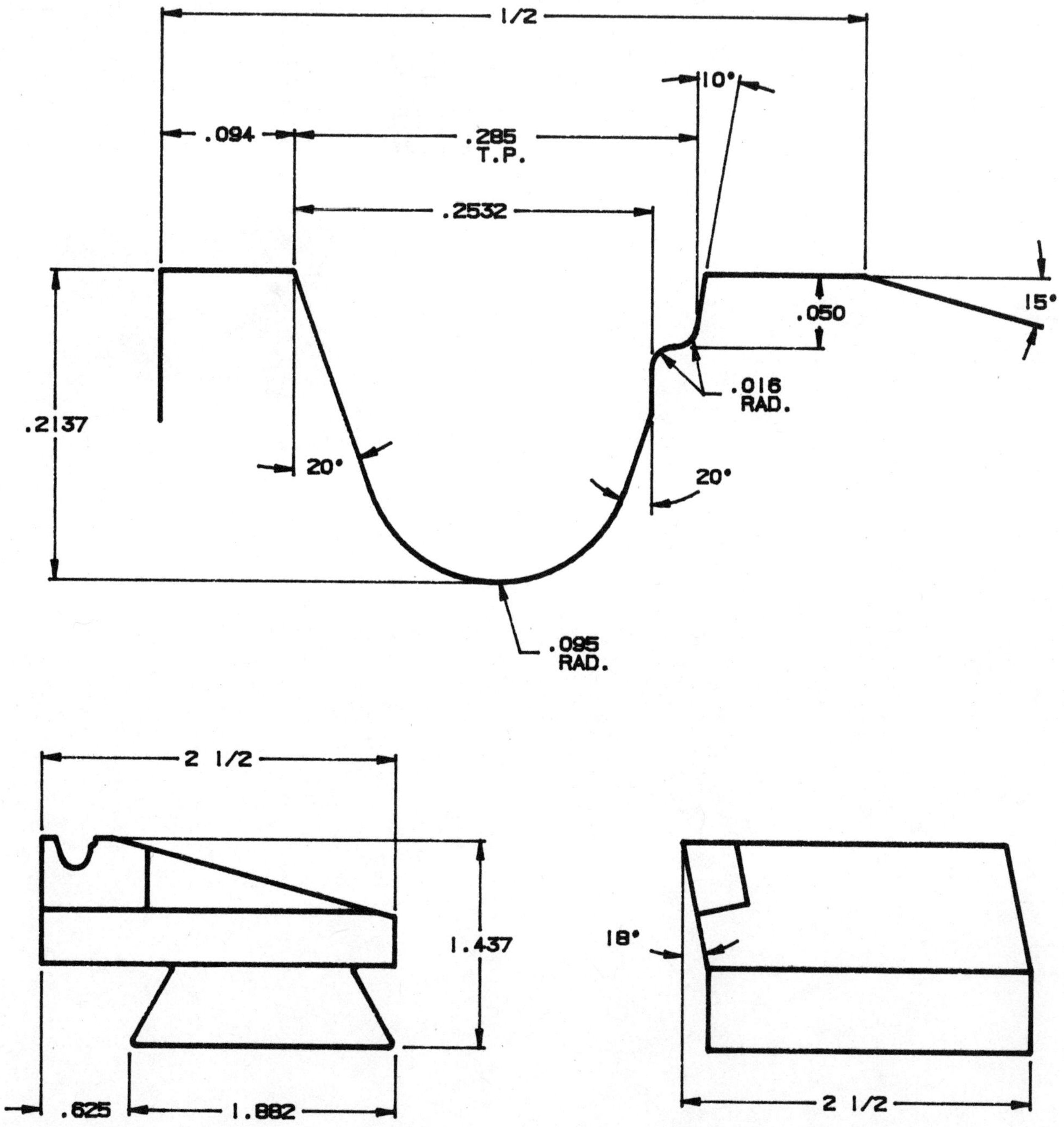

C-2 CARBIDE, T.S. HD'N. & GR'D.

DET.	REQ'D.	FINISH SIZE	MAT'L.

THE PYLE-NATIONAL COMPANY
CHICAGO, ILLINOIS 60651

T.C.T. DOVETAIL FORM TOOL

UNLESS OTHERWISE SHOWN TOLERANCES

FRACTIONAL ±1/64
DECIMAL ±.001
ANGULAR ±0°15'

DR.	DATE	DWG. NO.
CHKD.	APP.	

REV.	DATE	BY	CHANGES

COMPUTER-AIDED DRAFTING AND DESIGN	DRAWN BY	DATE	DWG NO.

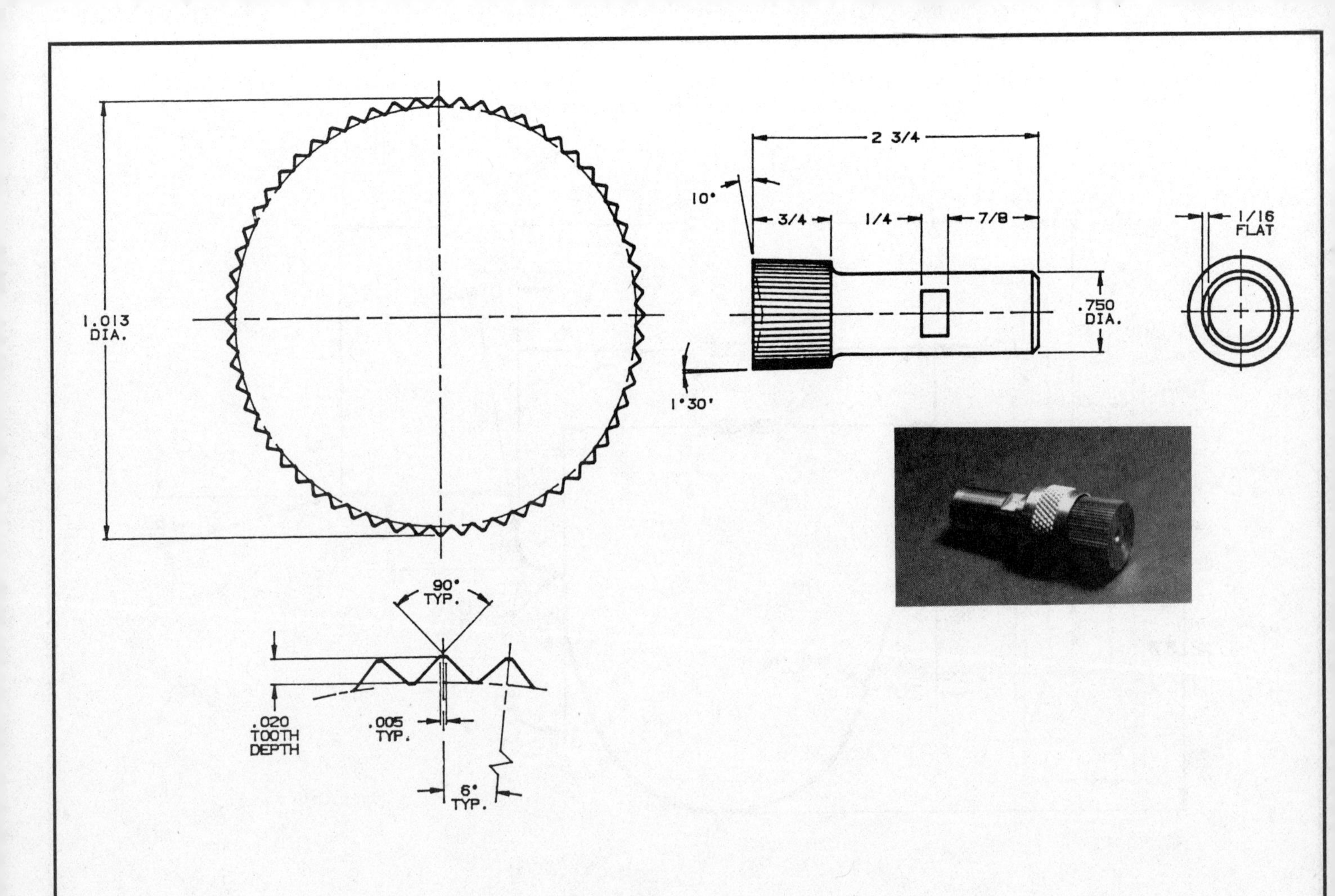

Develop necessary details to create a working drawing of the *Rotary Broach.* Change all fractions to decimals. Change the length of the cutting surface from .75 to .88 and make the overall length 2.88. Plot your drawing on the bottom half of this sheet or on a separate blank sheet.

COMPUTER-AIDED DRAFTING AND DESIGN	DRAWN BY	DATE	DWG NO.

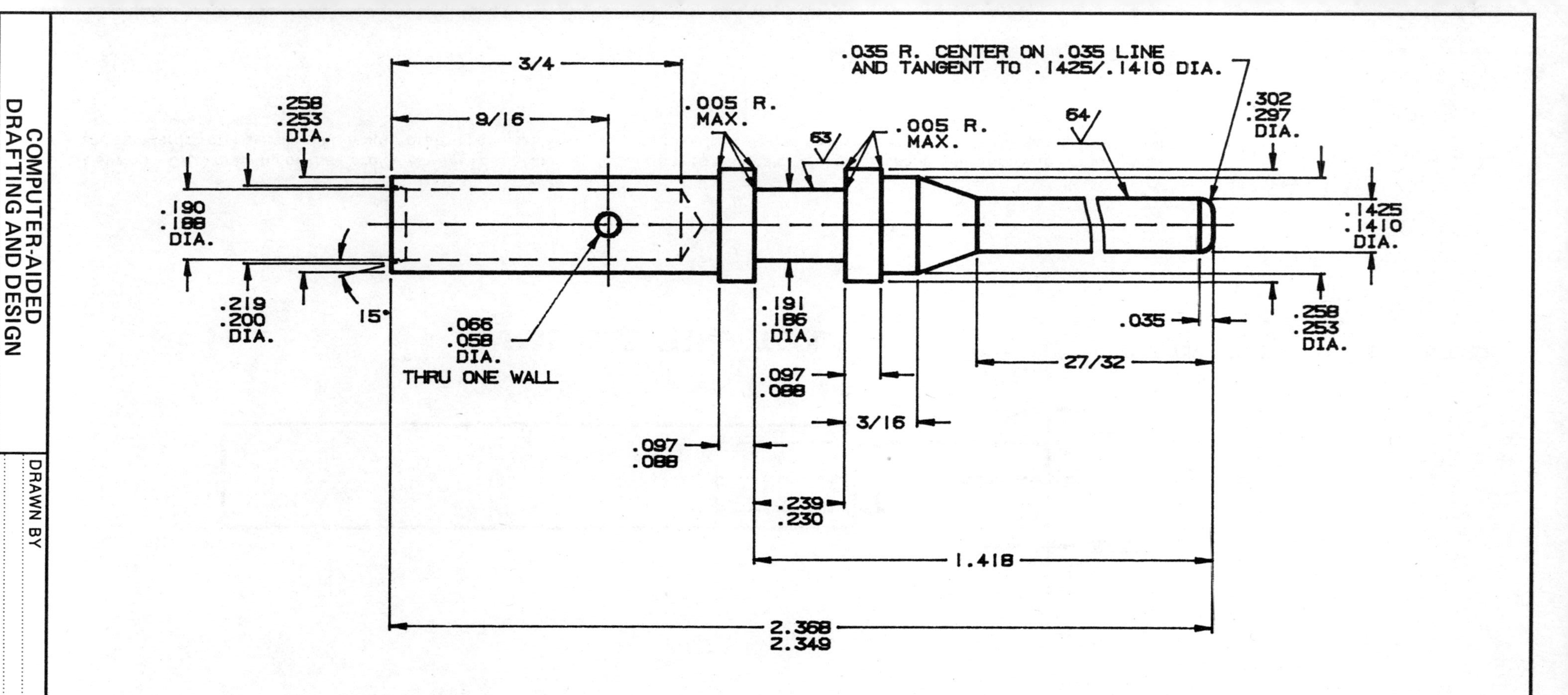

Develop a CAD drawing of the *Forming Tool.* Change all fractions to decimals. Make any changes necessary to comply with the standards you have learned in your drafting course. Plot your finished drawing on the bottom of this sheet or on a separate blank sheet.

COMPUTER-AIDED DRAFTING AND DESIGN	DRAWN BY	DATE	DWG NO.

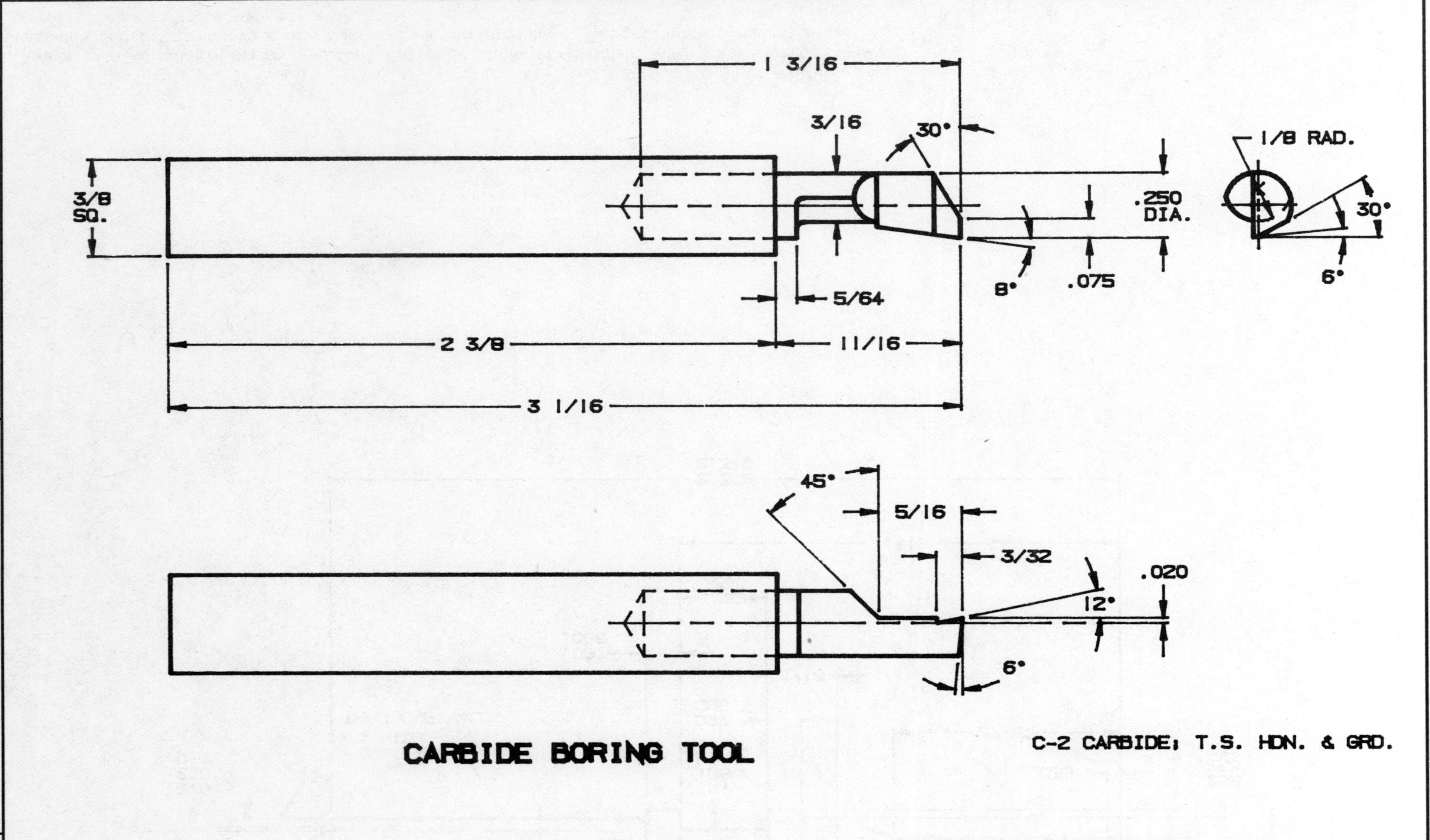

Develop a CAD drawing of the *Carbide Boring Tool.* Change all fractions to decimals. Make any changes necessary to comply with the standards you have learned in your drafting course. Plot your finished drawing on the reverse side of an earlier completed drawing or on a separate sheet.

COMPUTER-AIDED DRAFTING AND DESIGN	DRAWN BY	DATE	DWG NO.

Develop a CAD drawing of the *Forming Block.* Make any changes necessary to comply with the standards you have learned in your drafting course. Plot your finished drawing on the bottom half of this sheet.

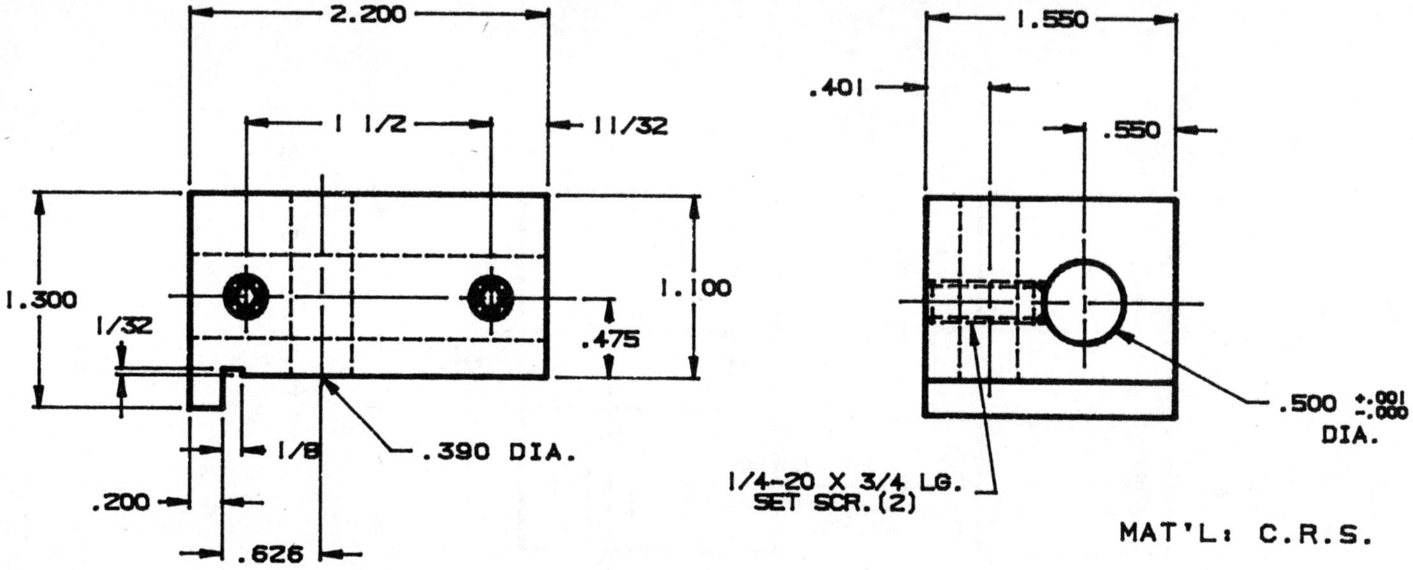

COMPUTER-AIDED DRAFTING AND DESIGN	DRAWN BY	DATE	DWG NO.

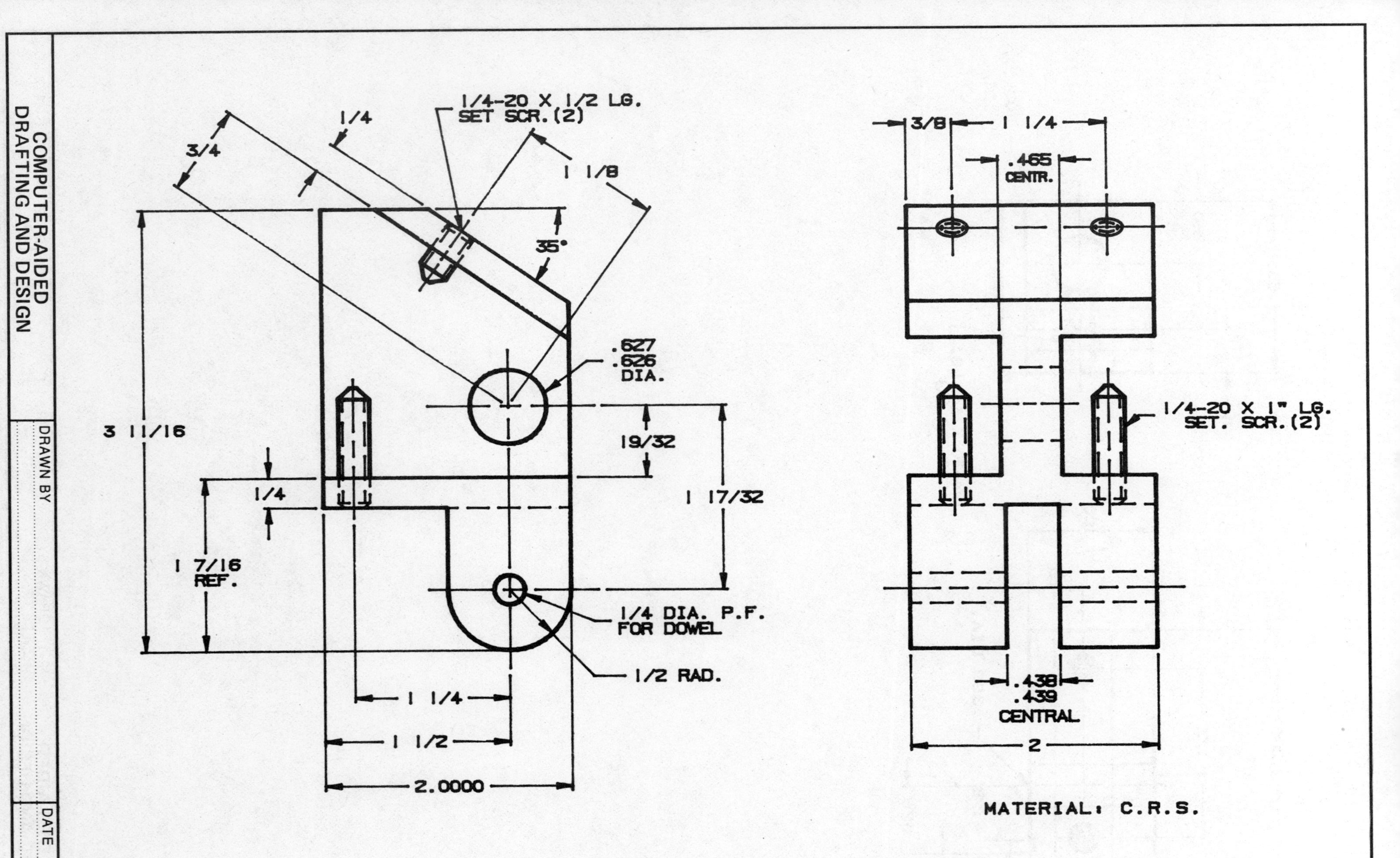

Create a detailed CAD drawing of the *Adjustable Inclined Plate.* Make any changes necessary to comply with the standards you have learned in your drafting course.

COMPUTER-AIDED DRAFTING AND DESIGN	DRAWN BY	DATE	DWG NO.

DRAWN BY
DATE
DWG NO.

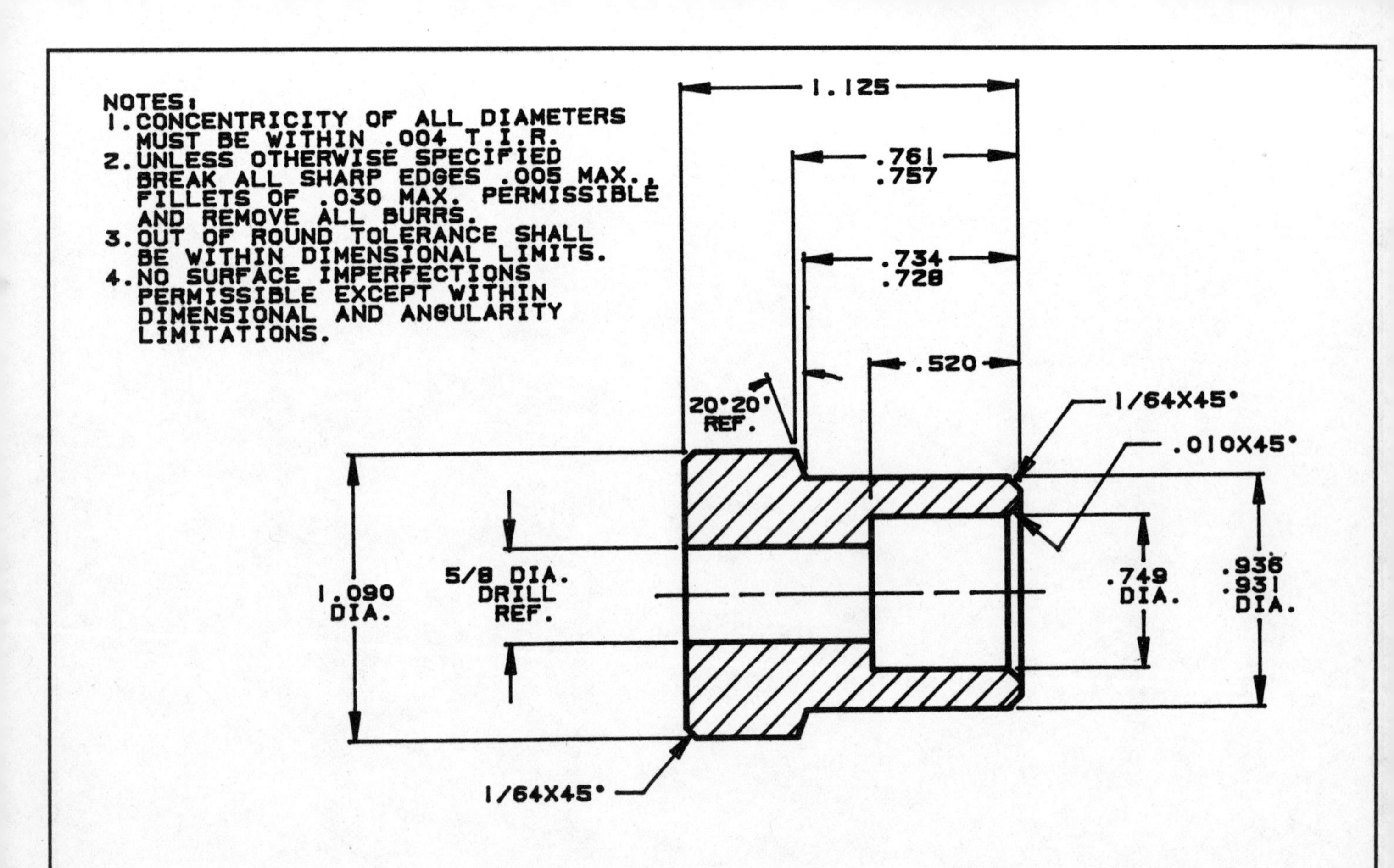

Develop a CAD drawing of the *Bearing Socket.* Make any changes necessary to comply with the standards you have learned in your drafting course. Plot your finished drawing on the bottom half of this sheet or on a separate sheet.

COMPUTER-AIDED DRAFTING AND DESIGN	DRAWN BY	DATE	DWG NO.

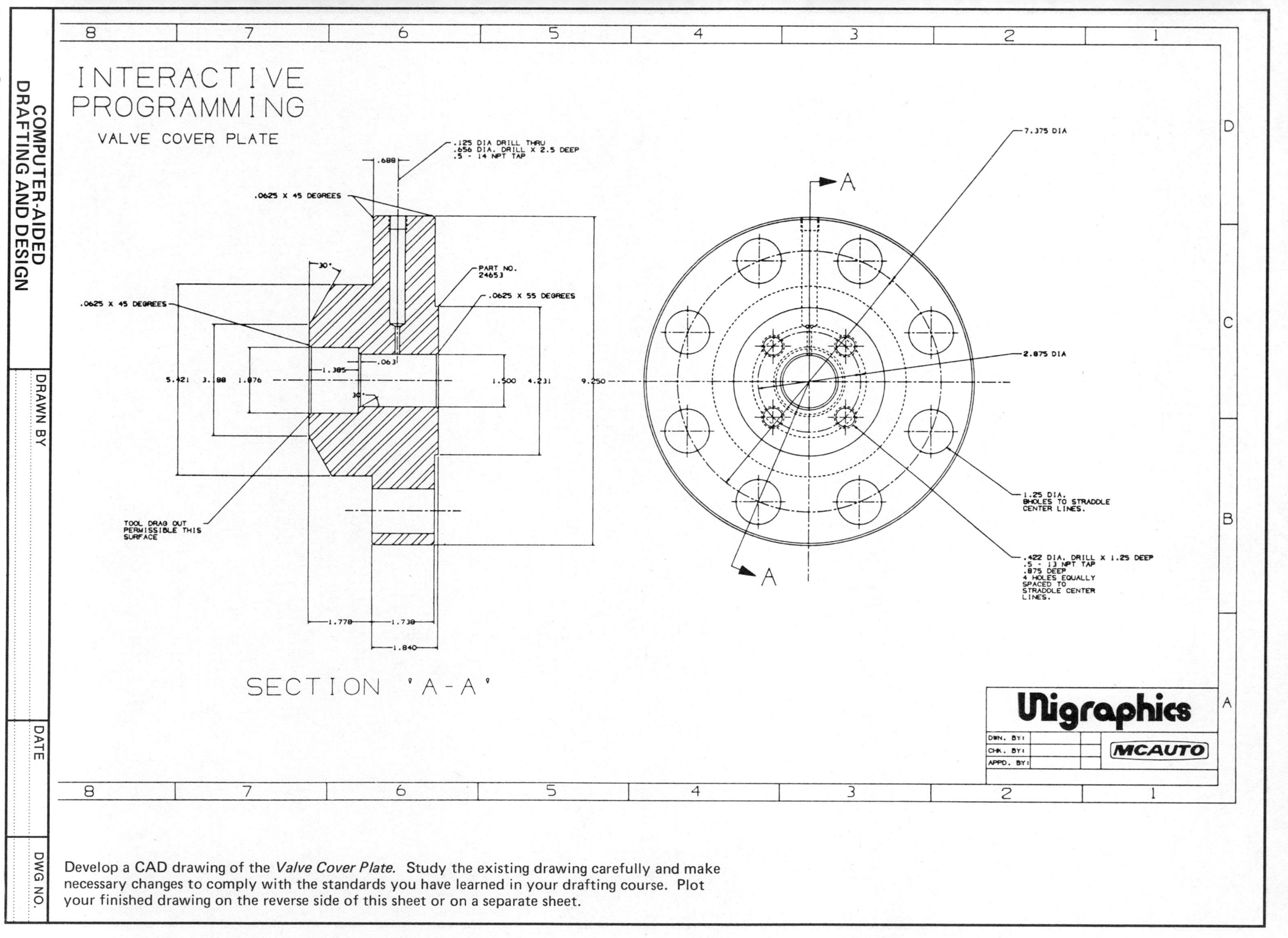

Develop a CAD drawing of the *Valve Cover Plate.* Study the existing drawing carefully and make necessary changes to comply with the standards you have learned in your drafting course. Plot your finished drawing on the reverse side of this sheet or on a separate sheet.

COMPUTER-AIDED DRAFTING AND DESIGN	DRAWN BY	DATE	DWG NO.

DRAWN BY
DATE
DWG NO.

Develop a CAD drawing of the *Adjustable Socket.* Convert fractions to decimals. Make any changes necessary to comply with the standards you have learned in your drafting course. Plot your finished drawing on the reverse side of this sheet or on a separate sheet.

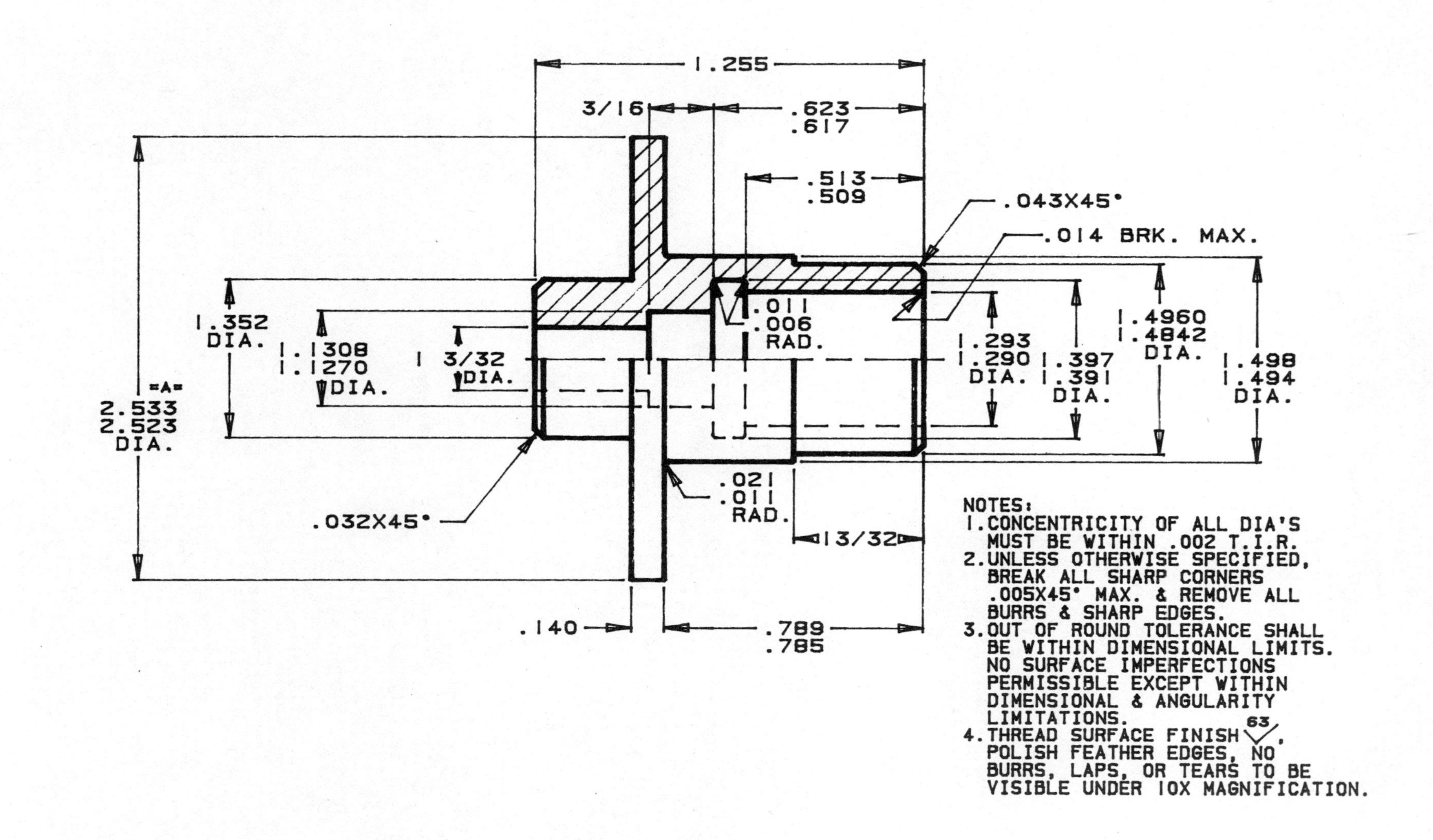

COMPUTER-AIDED DRAFTING AND DESIGN	DRAWN BY	DATE	DWG NO.

	DRAWN BY	DATE	DWG NO.

Develop a CAD drawing of the *Housing Insert.* Make necessary changes to comply with ANSI Standards.
Plot your finished drawing on the reverse side of this sheet or on a separate sheet.

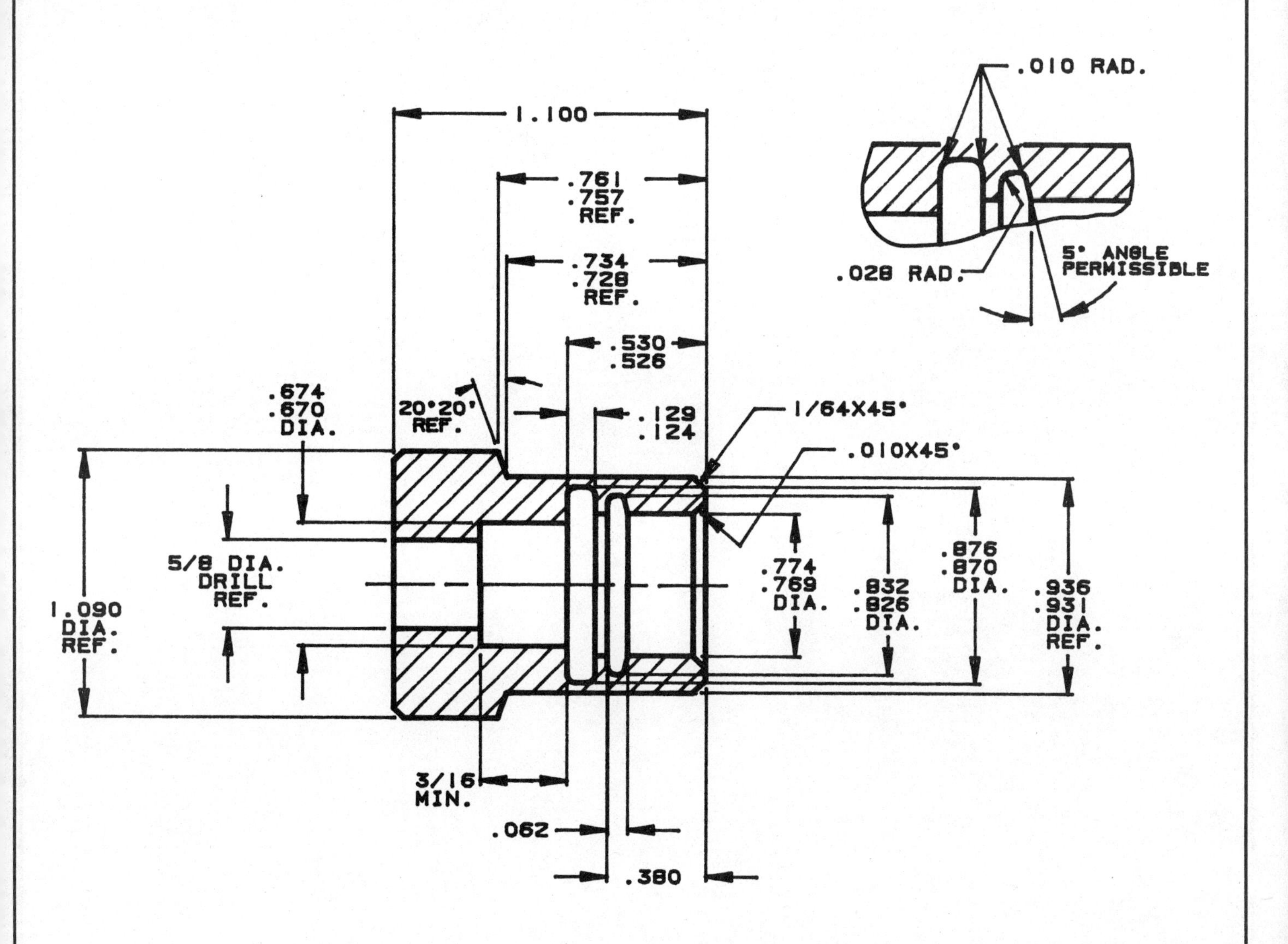

NOTES:
1. CONCENTRICITY OF ALL DIAMETERS MUST BE WITHIN .004 T.I.R.
2. UNLESS OTHERWISE SPECIFIED BREAK ALL SHARP EDGES .005 MAX., FILLETS OF .030 MAX. PERMISSIBLE AND REMOVE ALL BURRS.
3. OUT OF ROUND TOLERANCE SHALL BE WITHIN DIMENSIONAL LIMITS.
4. NO SURFACE IMPERFECTIONS PERMISSIBLE EXCEPT WITHIN DIMENSIONAL AND ANGULARITY LIMITATIONS.

COMPUTER-AIDED DRAFTING AND DESIGN	DRAWN BY	DATE	DWG NO.

DRAWN BY
DATE
DWG NO.

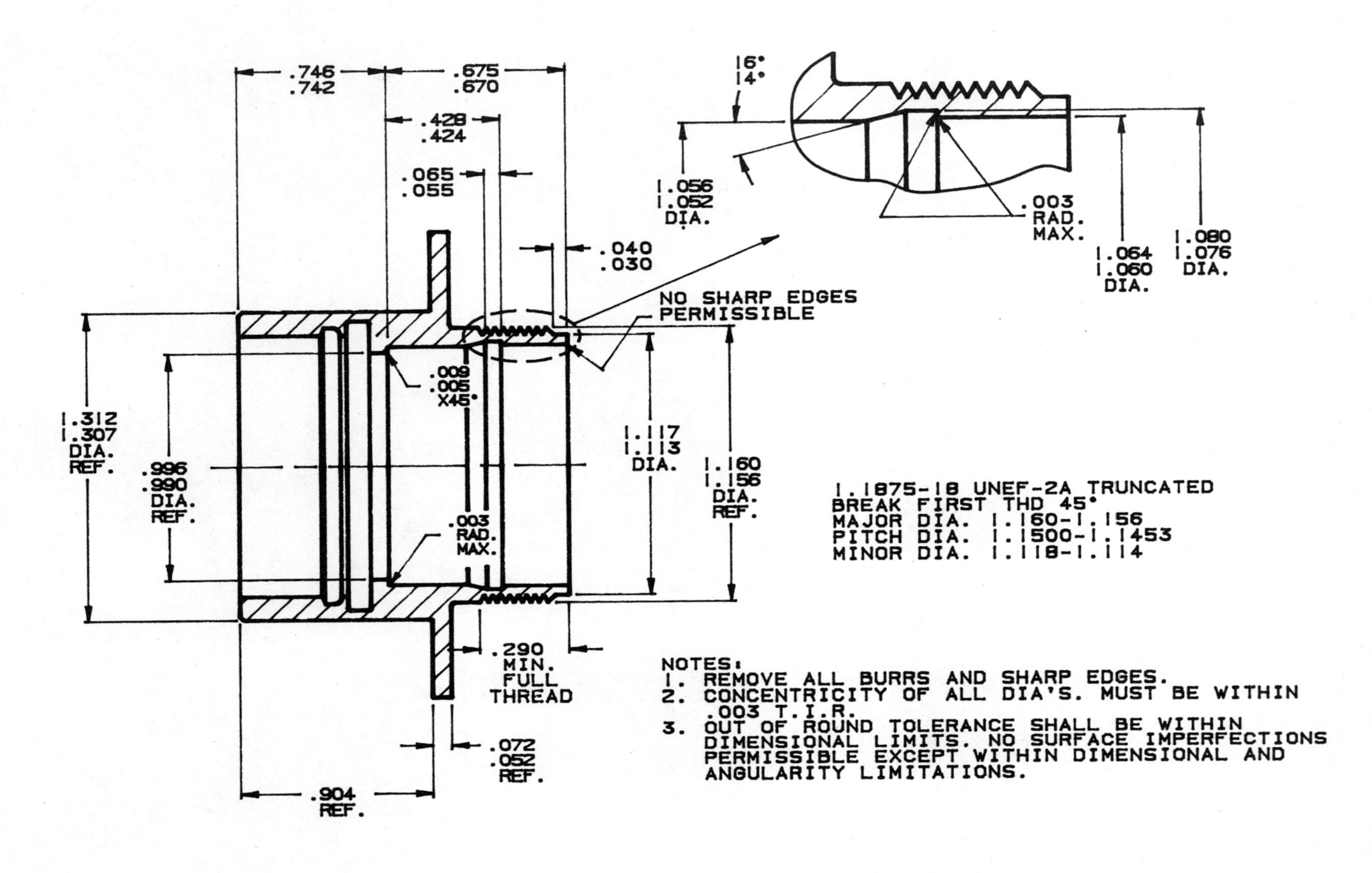

Develop a CAD drawing of the *Threaded Insert Housing.* Make necessary changes to comply with the standards you have learned in your drafting course. Plot your finished drawing on the reverse side of this sheet or on a separate sheet.

COMPUTER-AIDED DRAFTING AND DESIGN	DRAWN BY	DATE	DWG NO.

DRAWN BY
DATE
DWG NO.

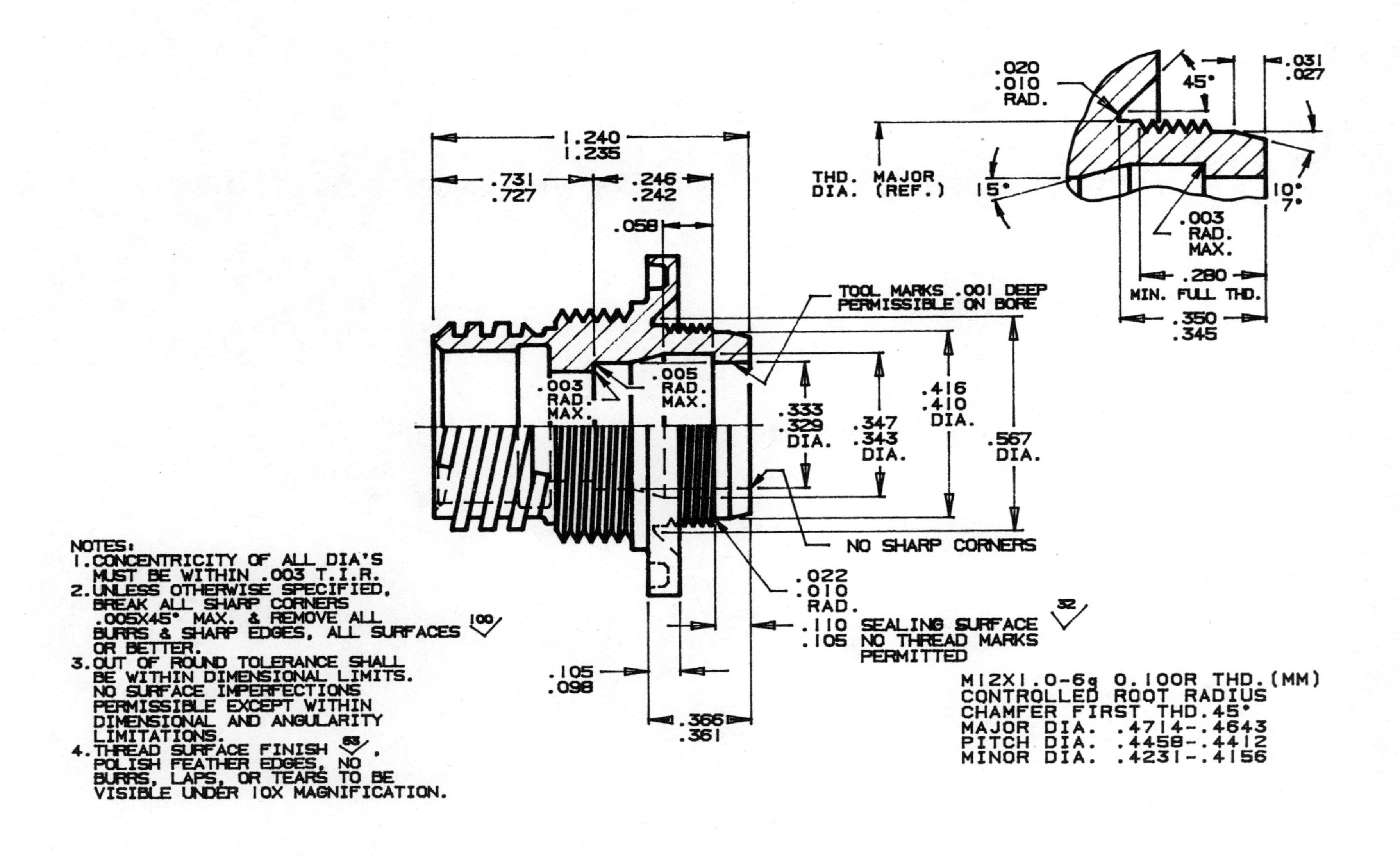

Develop a CAD drawing of the *Threaded Coupling.* Make all necessary changes to comply with the standards you have learned in your drafting course. Plot your finished drawing on the reverse side of this sheet or on a separate sheet.

COMPUTER-AIDED DRAFTING AND DESIGN	DRAWN BY	DATE	DWG NO.

	DRAWN BY	DATE	DWG NO.

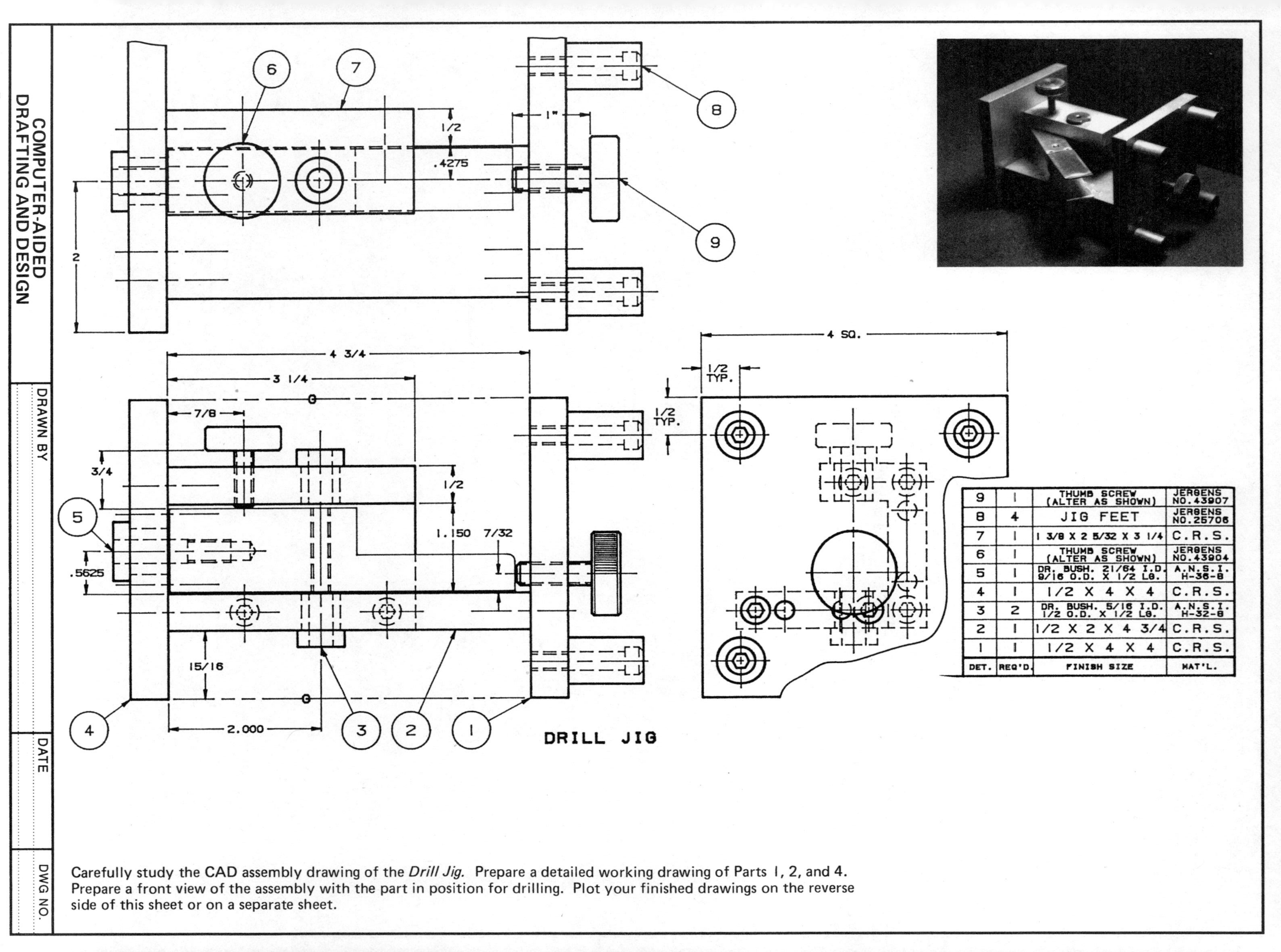

DET.	REQ'D.	FINISH SIZE	MAT'L.
9	1	THUMB SCREW (ALTER AS SHOWN)	JERGENS NO.43907
8	4	JIG FEET	JERGENS NO.25706
7	1	1 3/8 X 2 5/32 X 3 1/4	C.R.S.
6	1	THUMB SCREW (ALTER AS SHOWN)	JERGENS NO.43904
5	1	DR. BUSH. 21/64 I.D. 9/16 O.D. X 1/2 LG.	A.N.S.I. H-36-8
4	1	1/2 X 4 X 4	C.R.S.
3	2	DR. BUSH. 5/16 I.D. 1/2 O.D. X 1/2 LG.	A.N.S.I. H-32-8
2	1	1/2 X 2 X 4 3/4	C.R.S.
1	1	1/2 X 4 X 4	C.R.S.

Carefully study the CAD assembly drawing of the *Drill Jig*. Prepare a detailed working drawing of Parts 1, 2, and 4. Prepare a front view of the assembly with the part in position for drilling. Plot your finished drawings on the reverse side of this sheet or on a separate sheet.

COMPUTER-AIDED DRAFTING AND DESIGN

DRAWN BY	DATE	DWG NO.

	DRAWN BY	DATE	DWG NO.

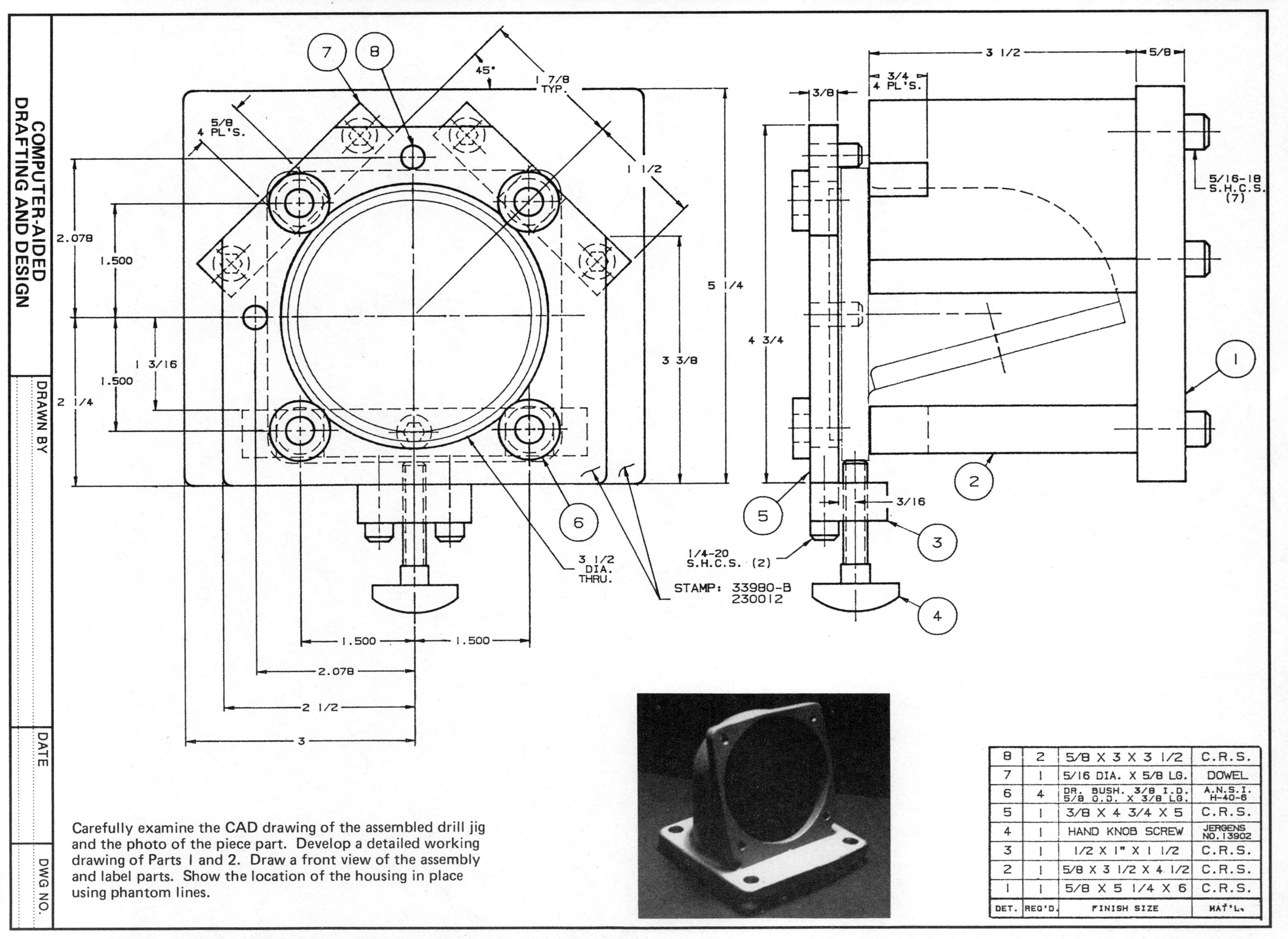

DET.	REQ'D.	FINISH SIZE	MAT'L.
8	2	5/8 X 3 X 3 1/2	C.R.S.
7	1	5/16 DIA. X 5/8 LG.	DOWEL
6	4	DR. BUSH. 3/8 I.D. 5/8 O.D. X 3/8 LG.	A.N.S.I. H-40-6
5	1	3/8 X 4 3/4 X 5	C.R.S.
4	1	HAND KNOB SCREW	JERGENS NO. 13902
3	1	1/2 X 1" X 1 1/2	C.R.S.
2	1	5/8 X 3 1/2 X 4 1/2	C.R.S.
1	1	5/8 X 5 1/4 X 6	C.R.S.

Carefully examine the CAD drawing of the assembled drill jig and the photo of the piece part. Develop a detailed working drawing of Parts 1 and 2. Draw a front view of the assembly and label parts. Show the location of the housing in place using phantom lines.

		DRAWN BY	DATE	DWG NO.

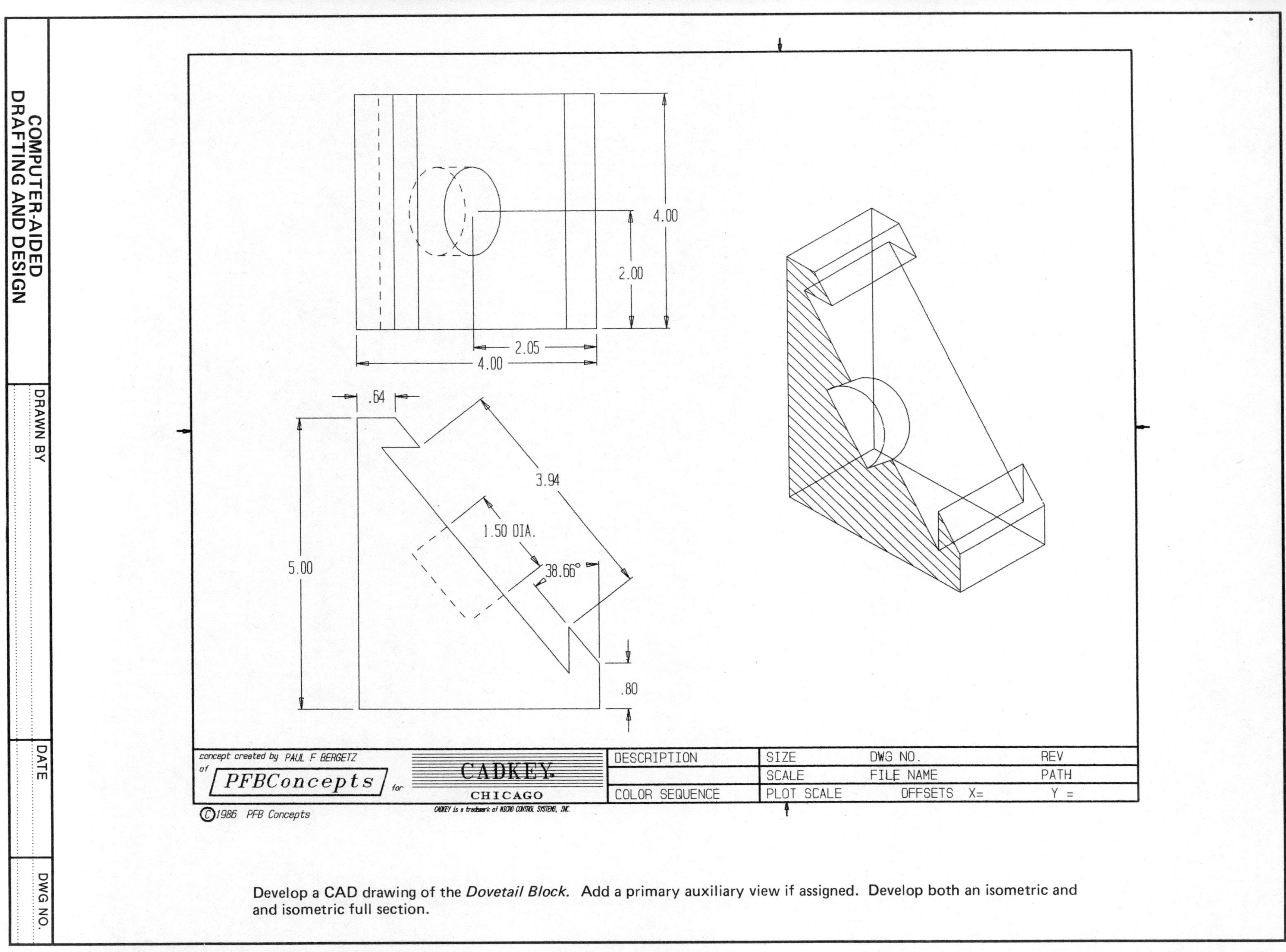

Develop a CAD drawing of the *Dovetail Block*. Add a primary auxiliary view if assigned. Develop both an isometric and and isometric full section.

	DRAWN BY	DATE	DWG NO.

Prepare a CAD drawing of the *Idler Plate.* Plot the finished drawing on the top half of this sheet. Notice the dimensioning method used. What is this method called? What purpose does it serve?

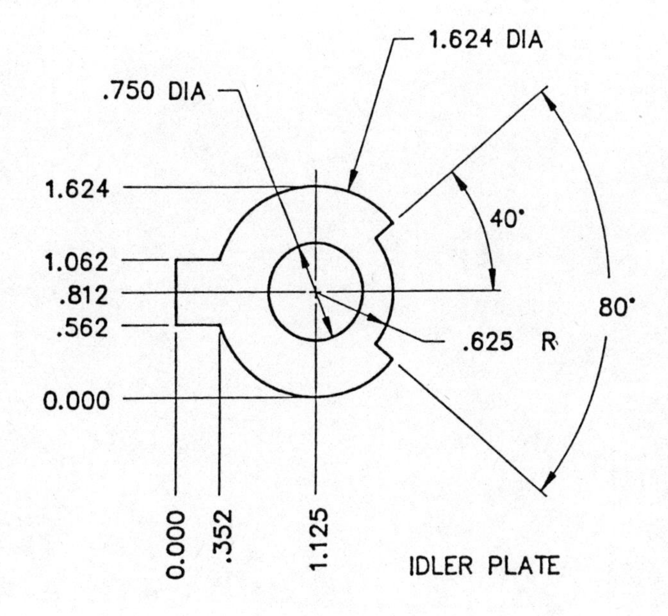

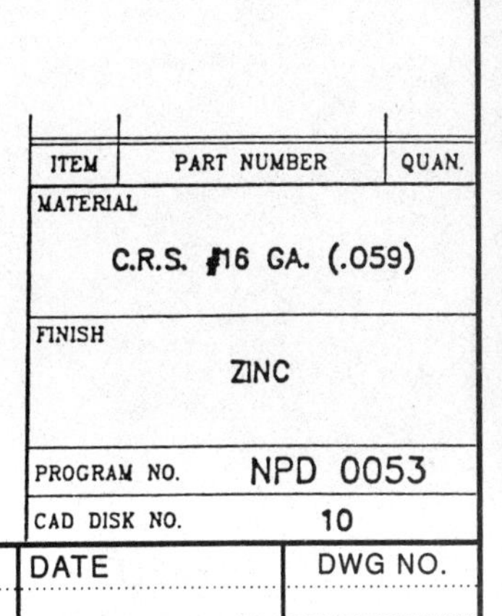

ITEM	PART NUMBER	QUAN.
MATERIAL	C.R.S. #16 GA. (.059)	
FINISH	ZINC	
PROGRAM NO.	NPD 0053	
CAD DISK NO.	10	

COMPUTER-AIDED DRAFTING AND DESIGN	DRAWN BY	DATE	DWG NO.

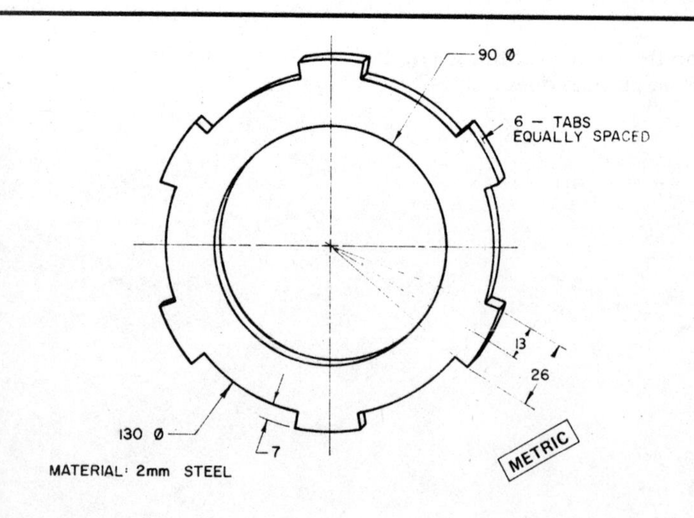

Prepare a CAD drawing of the *Clutch Disc.* Plot the finished drawing on this sheet. Make any changes necessary to comply with standards you have learned in your drafting course. Scale optional.

COMPUTER-AIDED DRAFTING AND DESIGN	DRAWN BY	DATE	DWG NO.

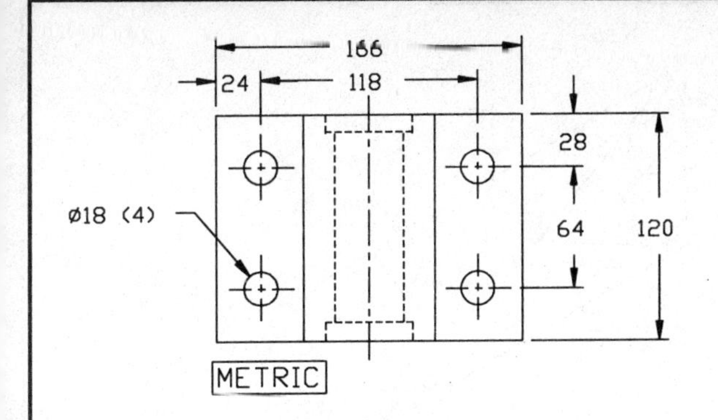

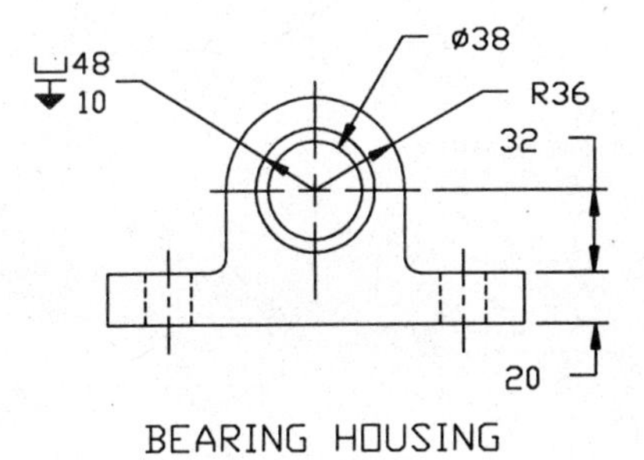

Prepare a CAD drawing of the *Bearing Housing.* Plot your finished drawing on the bottom of this sheet. Make any changes necessary to comply with standards you have learned in your drafting course. Scale optional.

COMPUTER-AIDED DRAFTING AND DESIGN	DRAWN BY	DATE	DWG NO.

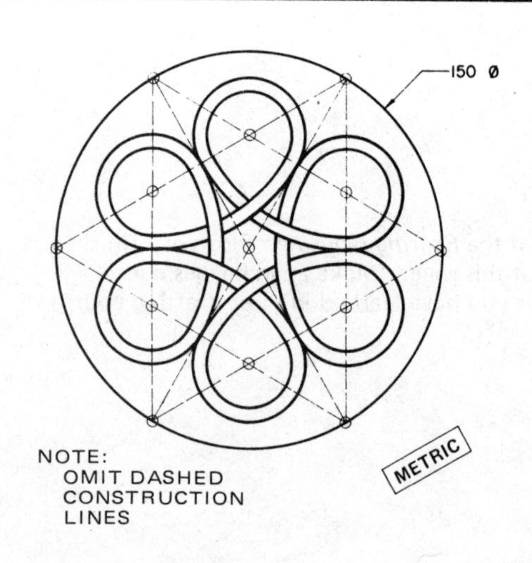

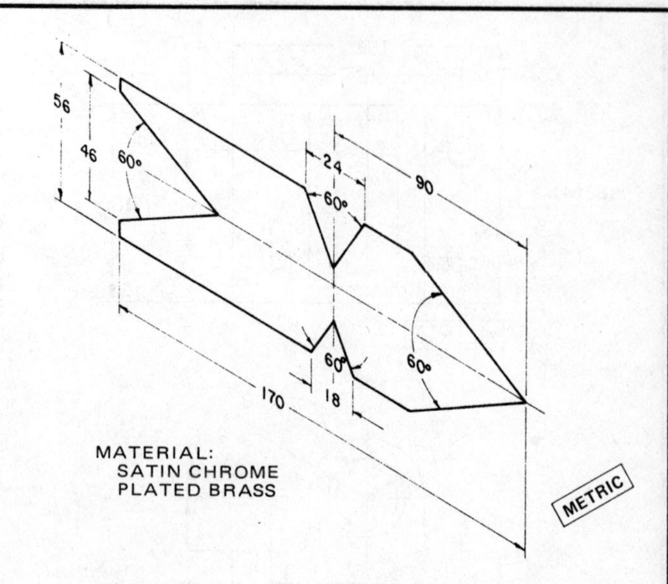

Prepare CAD drawings of the objects shown. Plot your finished drawings on this sheet or on separate sheets.

COMPUTER-AIDED DRAFTING AND DESIGN	DRAWN BY	DATE	DWG NO.

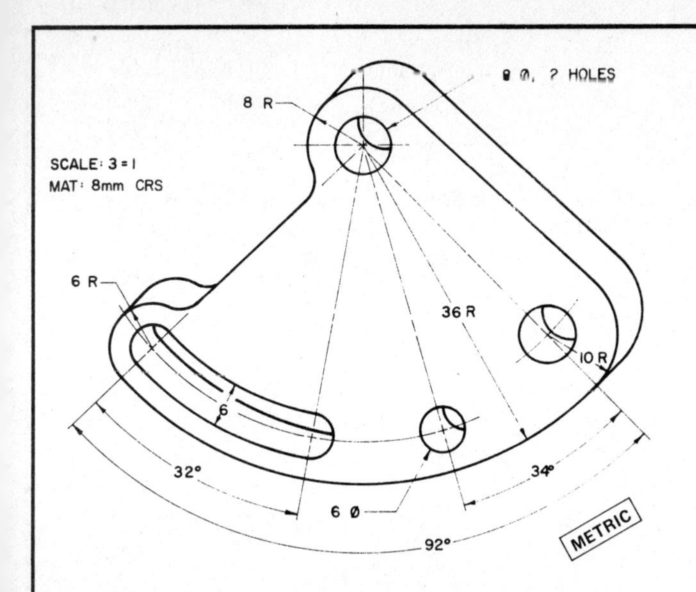

Prepare a CAD drawing of the *Adjusting Plate.* Plot your finished drawing on the bottom of this sheet. Scale optional. Convert metric sizes to U.S. Customary sizes if assigned.

COMPUTER-AIDED DRAFTING AND DESIGN	DRAWN BY	DATE	DWG NO.

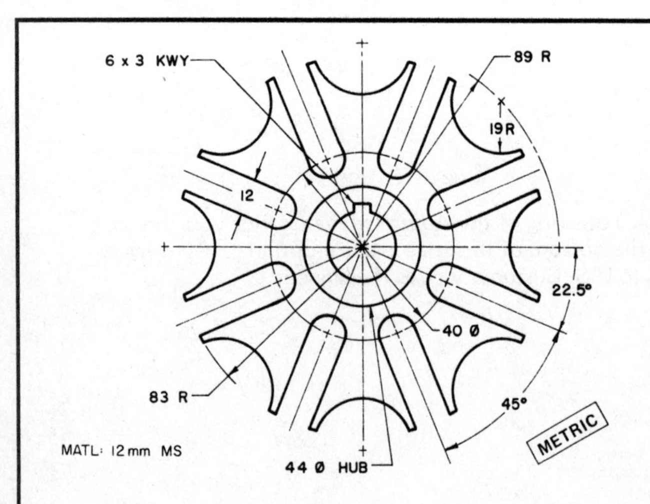

Prepare a CAD drawing of the *Sprocket.* Plot your finished drawing on the bottom of this sheet. Scale optional. Convert metric sizes to U. S. Customary sizes if assigned.

COMPUTER-AIDED DRAFTING AND DESIGN	DRAWN BY	DATE	DWG NO.

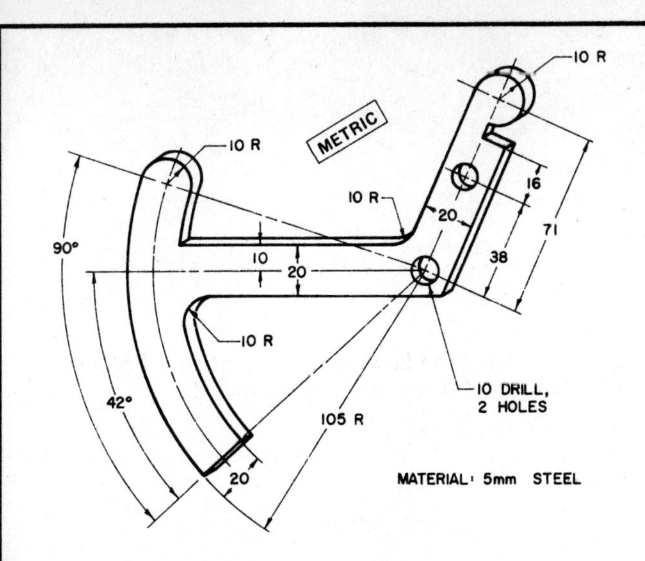

Prepare a CAD drawing of the *Rocker Arm.* Plot your finished drawing on the bottom of this sheet. Scale optional.

COMPUTER-AIDED DRAFTING AND DESIGN	DRAWN BY	DATE	DWG NO.

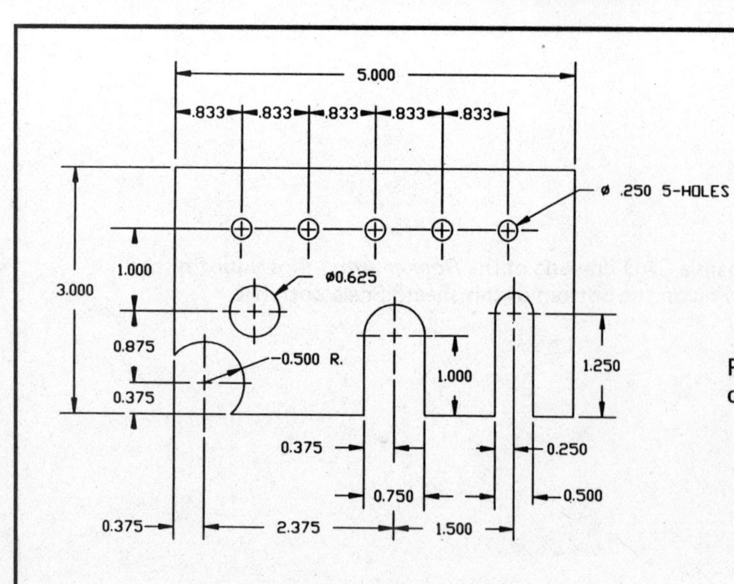

Prepare a CAD drawing of the *Locator Plate.* Plot your finished drawing on the bottom of this sheet. Scale optional.

COMPUTER-AIDED DRAFTING AND DESIGN	DRAWN BY	DATE	DWG NO.

Prepare a CAD drawing of the *Fan Guard.* Plot your finished drawing on a separate blank sheet or on the reverse side of this sheet. Make any changes necessary to comply with standards you have learned in your drafting course. Scale optional.

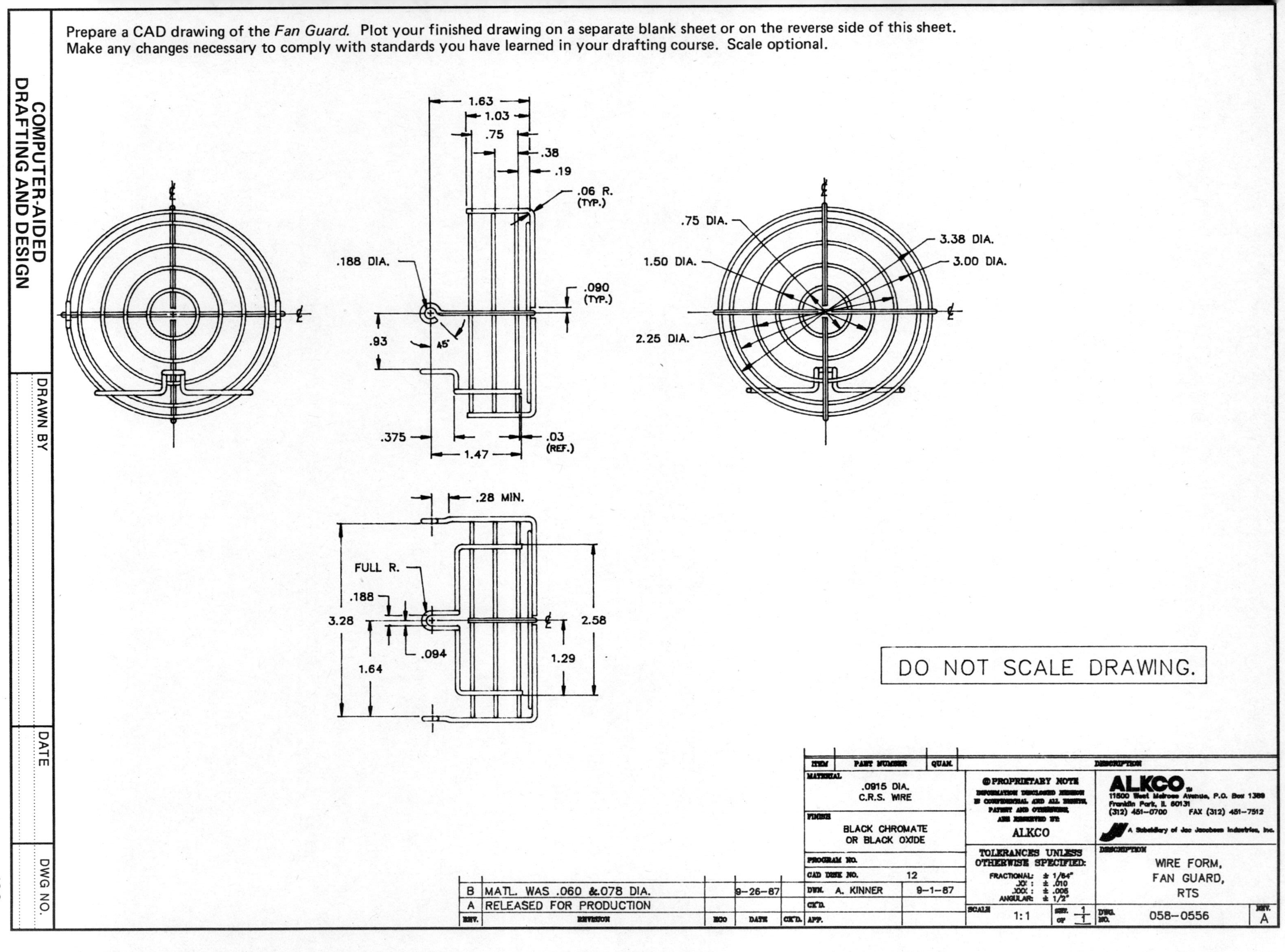

COMPUTER-AIDED DRAFTING AND DESIGN

DRAWN BY

DATE

DWG NO.

	DRAWN BY	DATE	DWG NO.

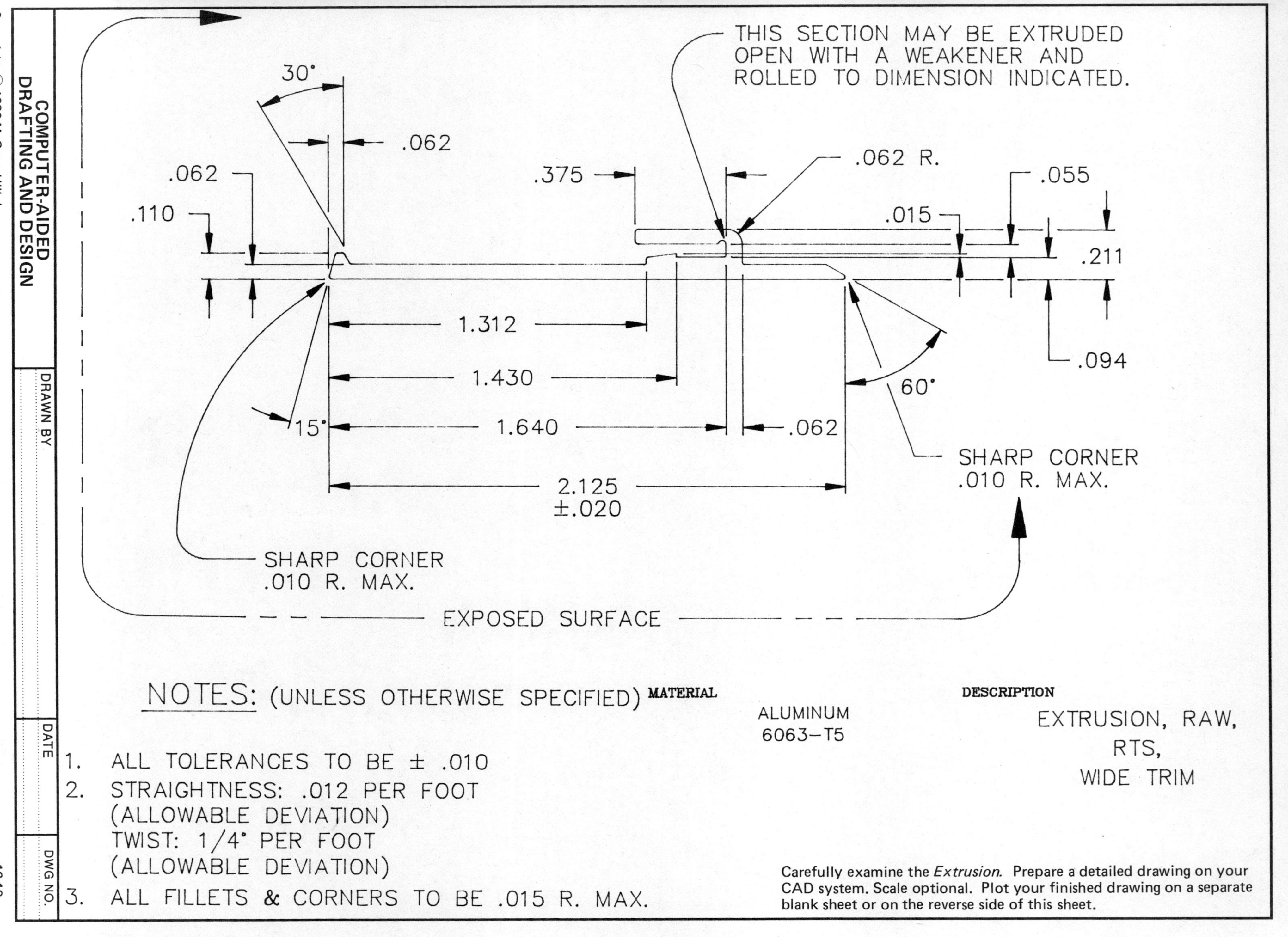
THIS SECTION MAY BE EXTRUDED OPEN WITH A WEAKENER AND ROLLED TO DIMENSION INDICATED.
30°
.062
.062
.110
.375
.062 R.
.055
.015
.211
.094
1.312
1.430
1.640
.062
15°
60°
2.125 ±.020
SHARP CORNER .010 R. MAX.
SHARP CORNER .010 R. MAX.
EXPOSED SURFACE
NOTES: (UNLESS OTHERWISE SPECIFIED)
1. ALL TOLERANCES TO BE ± .010
2. STRAIGHTNESS: .012 PER FOOT (ALLOWABLE DEVIATION)
TWIST: 1/4° PER FOOT (ALLOWABLE DEVIATION)
3. ALL FILLETS & CORNERS TO BE .015 R. MAX.
MATERIAL
ALUMINUM 6063–T5
DESCRIPTION
EXTRUSION, RAW, RTS, WIDE TRIM
Carefully examine the Extrusion. Prepare a detailed drawing on your CAD system. Scale optional. Plot your finished drawing on a separate blank sheet or on the reverse side of this sheet.
COMPUTER-AIDED DRAFTING AND DESIGN
DRAWN BY
DATE
DWG NO.

	DRAWN BY	DATE	DWG NO.

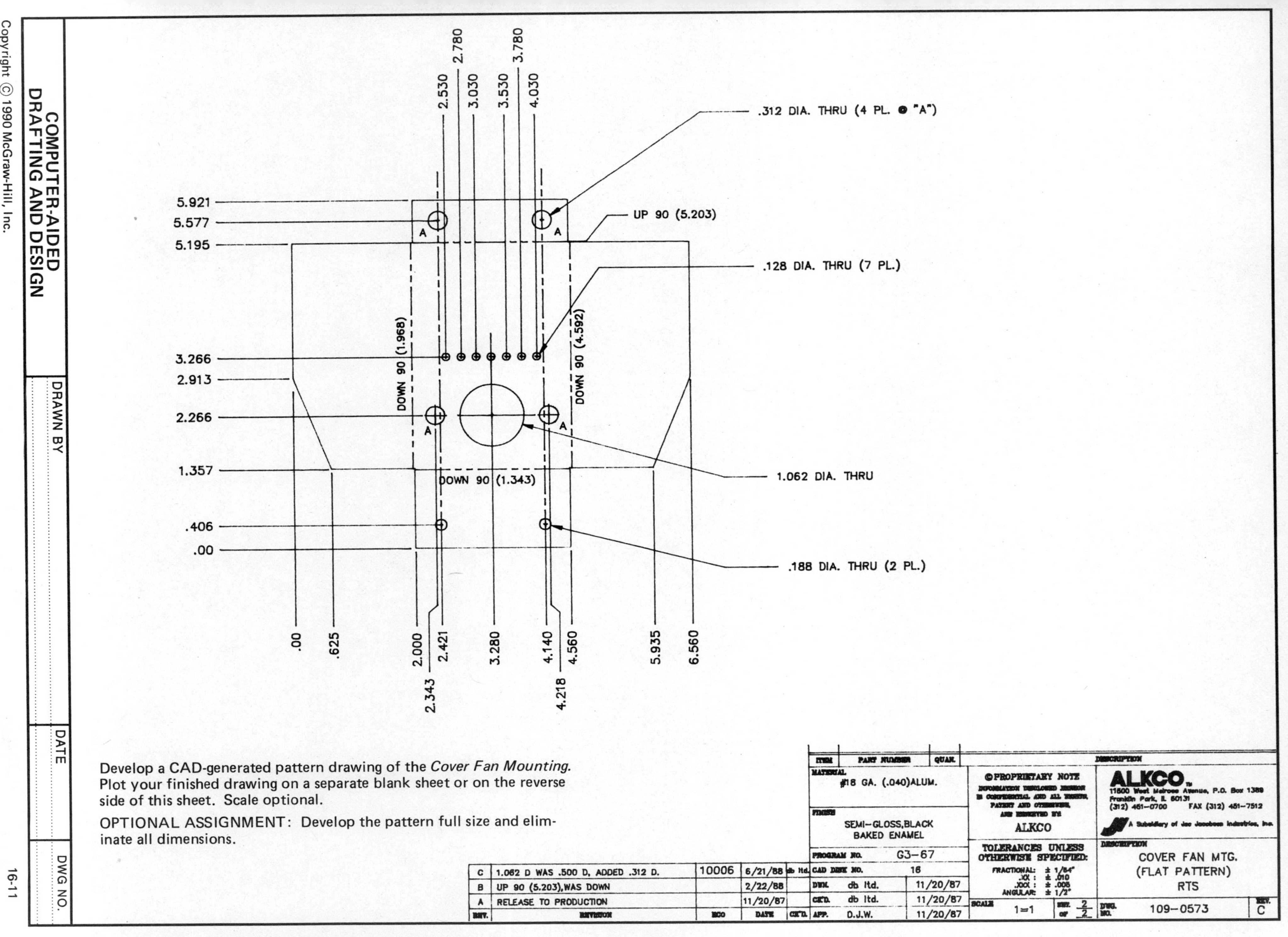

Develop a CAD-generated pattern drawing of the *Cover Fan Mounting.* Plot your finished drawing on a separate blank sheet or on the reverse side of this sheet. Scale optional.

OPTIONAL ASSIGNMENT: Develop the pattern full size and eliminate all dimensions.

COMPUTER-AIDED DRAFTING AND DESIGN

DRAWN BY	DATE	DWG NO.

	DRAWN BY	DATE	DWG NO.

NOTES: UNLESS OTHERWISE SPECIFIED

1– ALL INSIDE DRAFT ANGLES ARE TO BE 1–1/2°, OUTSIDE 1°.

2– TOLERANCES ARE TO BE ±.010, ANGLES ±1/2°.

3– TYPICAL WALL THICKNESSES ARE TO BE .090 (+ 1° DRAFT).

4– ALL EGDE, CORNER & FILLET RADII ARE TO BE .055 R. MAX.

5– CASTING SHALL BE SMOOTH, CLEAN, FREE OF HARMFUL POROSITY, CRACKS, INCLUSIONS, CHILLS, GATE MARKS OR ANY OTHER DEFECTS DETRIMENTAL TO MACHINABILITY OR PERFORMANCE.

6– EXTERIOR SURFACES ARE TO BE FREE OF FLASH, MOLD MARKS AND EJECTOR PIN MARKS.

7– EJECTOR PINS TO BE LOCATED ON INTERIOR SURFACES ONLY.

8– EJECTOR PINS TO BE FLUSH TO .015 (0.38 MM) BELOW SURFACE.

9– THIS PART TO MATE WITH PART NO. 068–0094.

10– FINISH: SEMI–GLOSS, HIGH TEMP. (475° F), BLACK BAKED ENAMEL.

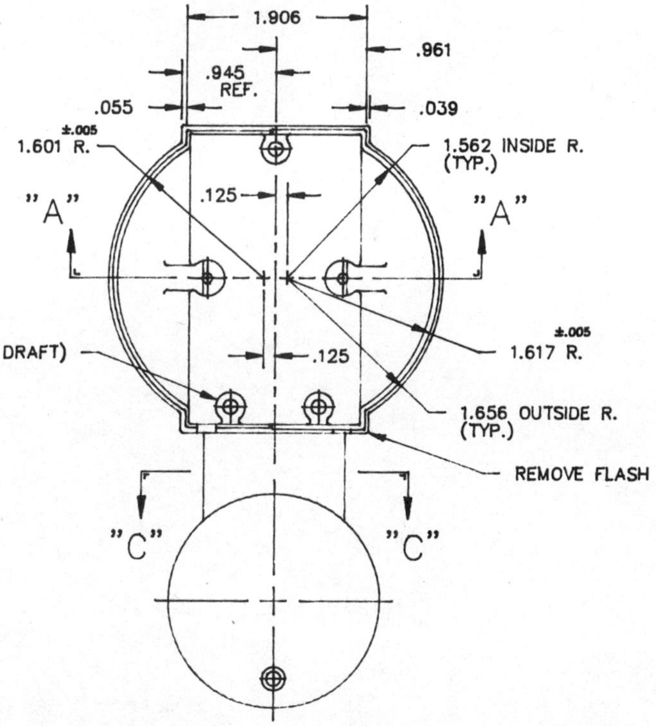

1.906
.961
.945 REF.
.055
.039
±.005 1.601 R.
1.562 INSIDE R. (TYP.)
.125
"A"
"A"
.156 R. (+ DRAFT) (3 PLCS)
.125
±.005 1.617 R.
1.656 OUTSIDE R. (TYP.)
REMOVE FLASH
"C"
"C"

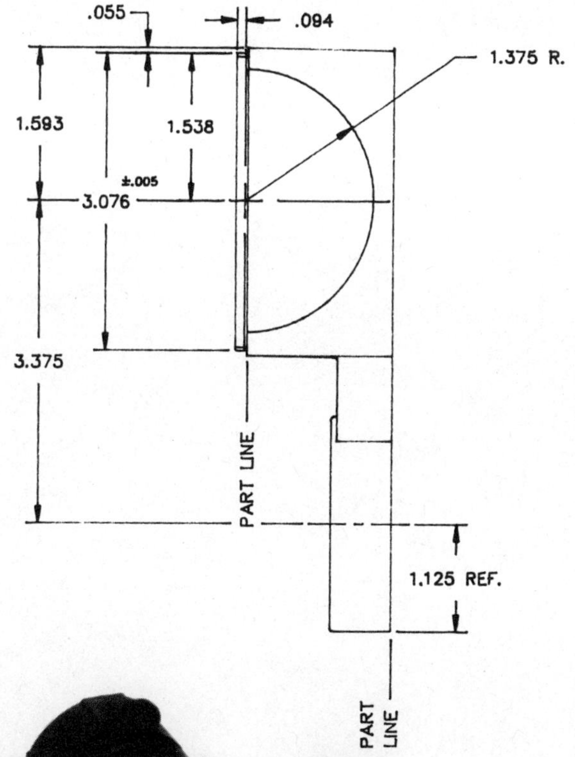

.055
.094
1.375 R.
1.593
1.538
±.005 3.076
3.375
PART LINE
1.125 REF.
PART LINE

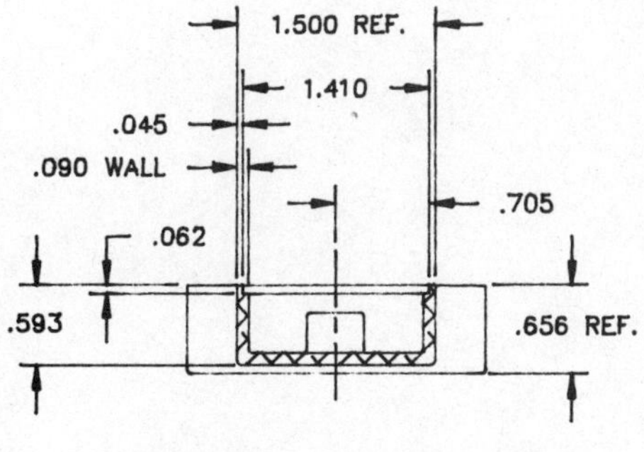

1.500 REF.
1.410
.045
.090 WALL
.705
.062
.593
.656 REF.

SECTION "C"–"C"

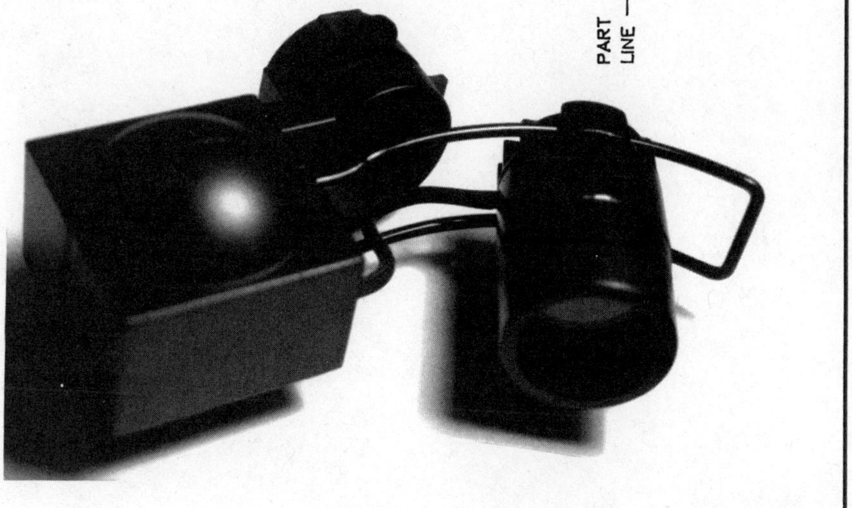

Use your CAD system to create the details essential for the manufacture of the *Light Module Component.* Study Sheets 16-12 and 16-13 before beginning.

G	REDRAWN ON CAD	88–5084	6/88	
REV.	REVISION	ECO	DATE	CK'D.

COMPUTER-AIDED DRAFTING AND DESIGN	DRAWN BY	DATE	DWG NO.

	DRAWN BY	DATE	DWG NO.

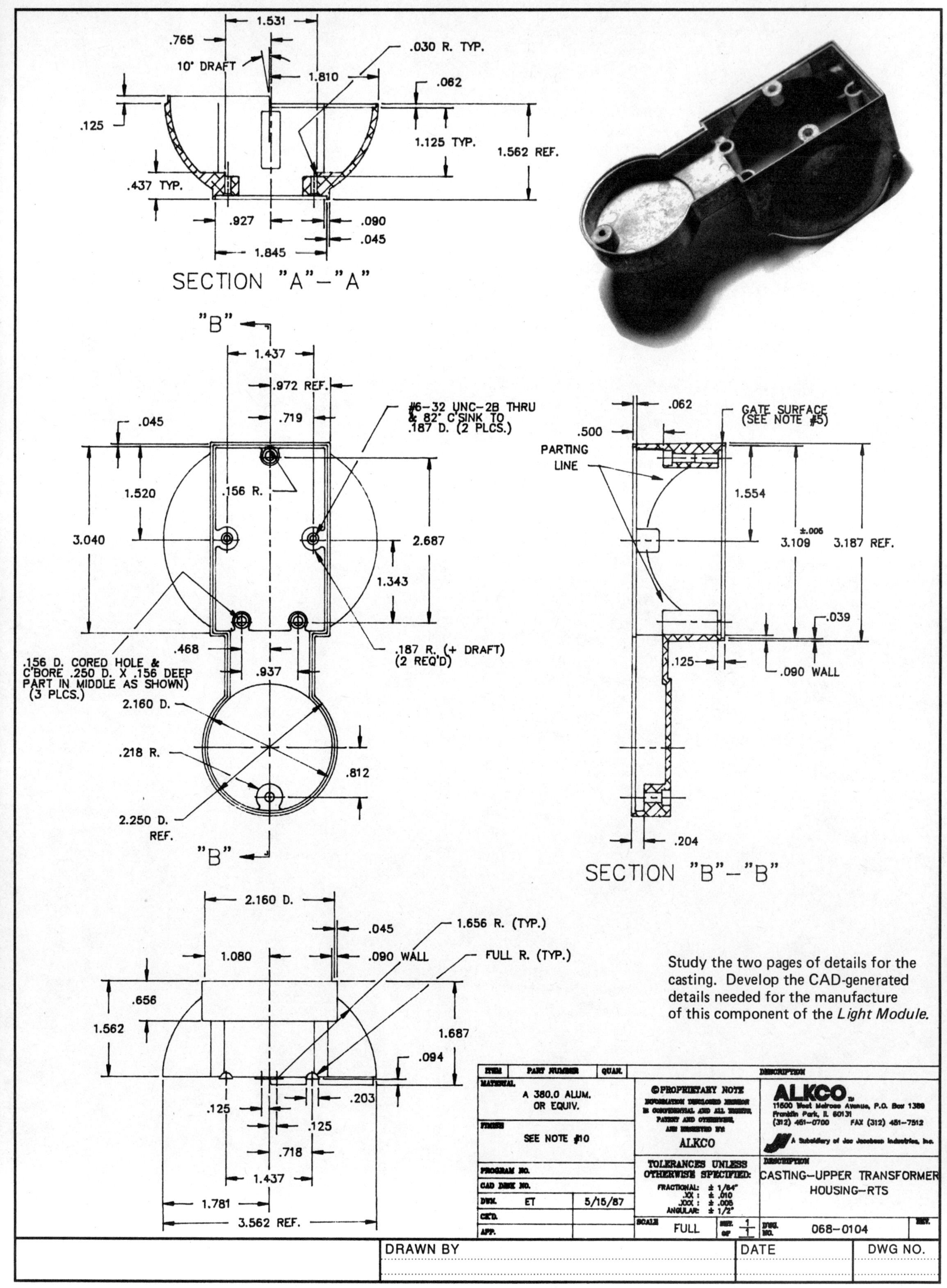

Study the two pages of details for the casting. Develop the CAD-generated details needed for the manufacture of this component of the *Light Module.*

DRAWN BY
DATE
DWG NO.

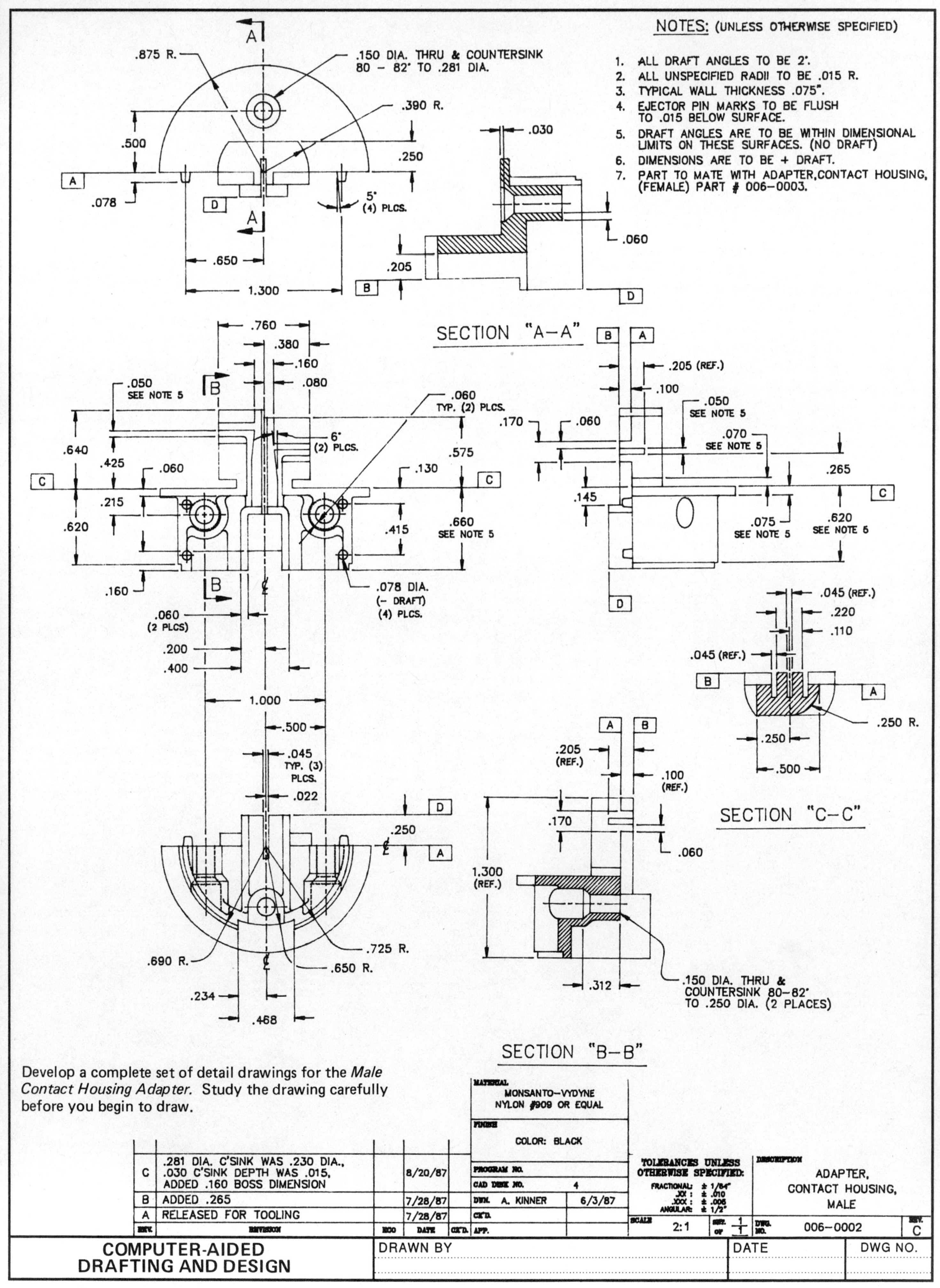
NOTES: (UNLESS OTHERWISE SPECIFIED)
1. ALL DRAFT ANGLES TO BE 2°.
2. ALL UNSPECIFIED RADII TO BE .015 R.
3. TYPICAL WALL THICKNESS .075".
4. EJECTOR PIN MARKS TO BE FLUSH TO .015 BELOW SURFACE.
5. DRAFT ANGLES ARE TO BE WITHIN DIMENSIONAL LIMITS ON THESE SURFACES. (NO DRAFT)
6. DIMENSIONS ARE TO BE + DRAFT.
7. PART TO MATE WITH ADAPTER,CONTACT HOUSING, (FEMALE) PART # 006-0003.
.875 R.
.150 DIA. THRU & COUNTERSINK 80 - 82° TO .281 DIA.
.390 R.
.500
.250
.078
5° (4) PLCS.
.650
1.300
.030
.060
.205
SECTION "A-A"
.760
.380
.160
.080
.050 SEE NOTE 5
.060 TYP. (2) PLCS.
6° (2) PLCS.
.640
.425
.060
.575
.130
.215
.620
.415
.660 SEE NOTE 5
.160
.078 DIA. (- DRAFT) (4) PLCS.
.060 (2 PLCS)
.200
.400
.205 (REF.)
.100
.170
.050 SEE NOTE 5
.070 SEE NOTE 5
.265
.145
.075 SEE NOTE 5
.620 SEE NOTE 5
.045 (REF.)
.220
.110
.250 R.
.250
.500
SECTION "C-C"
1.000
.500
.045 TYP. (3) PLCS.
.022
.250
.690 R.
.725 R.
.650 R.
.234
.468
.205 (REF.)
.100 (REF.)
.170
.060
1.300 (REF.)
.312
.150 DIA. THRU & COUNTERSINK 80-82° TO .250 DIA. (2 PLACES)
SECTION "B-B"
Develop a complete set of detail drawings for the Male Contact Housing Adapter. Study the drawing carefully before you begin to draw.
MATERIAL MONSANTO-VYDYNE NYLON #909 OR EQUAL
FINISH COLOR: BLACK
C .281 DIA. C'SINK WAS .230 DIA., .030 C'SINK DEPTH WAS .015, ADDED .160 BOSS DIMENSION 8/20/87
B ADDED .265 7/28/87
A RELEASED FOR TOOLING 7/28/87
REV. REVISION ECO DATE CK'D. APP.
PROGRAM NO.
CAD DISK NO. 4
DWN. A. KINNER 6/3/87
CK'D.
TOLERANCES UNLESS OTHERWISE SPECIFIED:
FRACTIONAL: ± 1/64"
.XX : ± .010
.XXX : ± .005
ANGULAR: ± 1/2°
DESCRIPTION ADAPTER, CONTACT HOUSING, MALE
SCALE 2:1
SHT. 1 OF 1
DWG. NO. 006-0002
REV. C
COMPUTER-AIDED DRAFTING AND DESIGN
DRAWN BY
DATE
DWG NO.

	DRAWN BY	DATE	DWG NO.

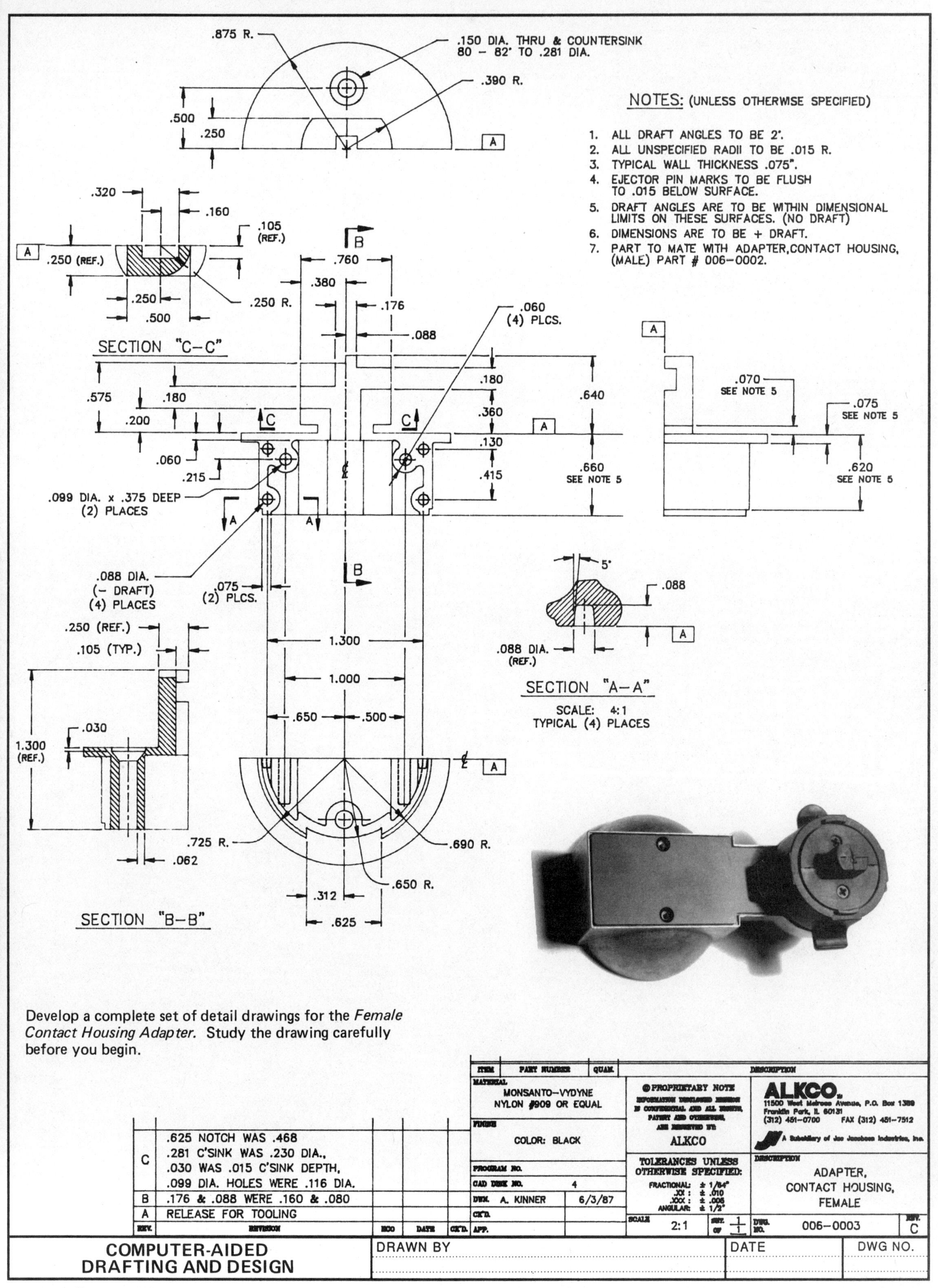

Develop a complete set of detail drawings for the *Female Contact Housing Adapter.* Study the drawing carefully before you begin.

	DRAWN BY	DATE	DWG NO.

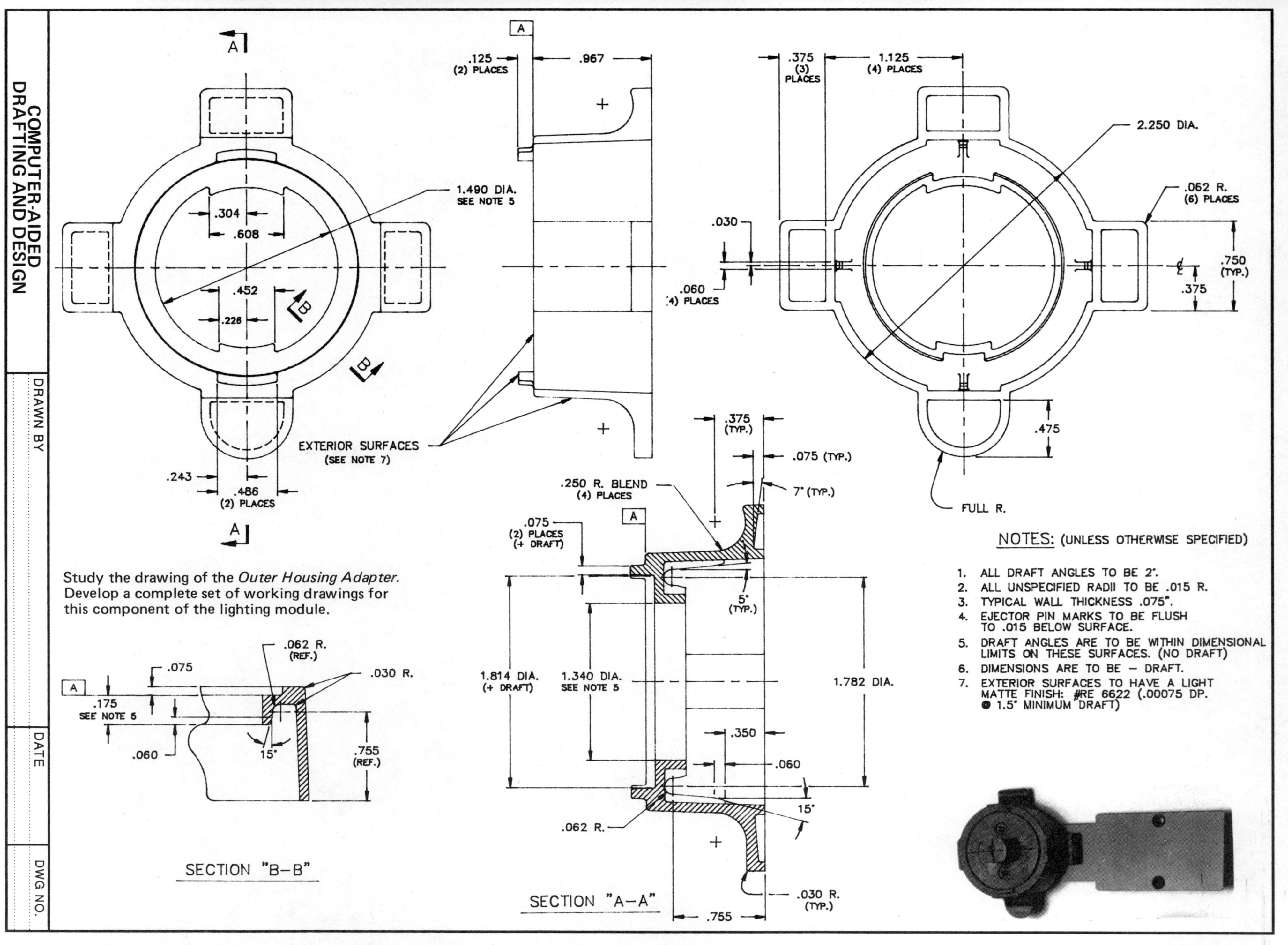

Study the drawing of the *Outer Housing Adapter.* Develop a complete set of working drawings for this component of the lighting module.

NOTES: (UNLESS OTHERWISE SPECIFIED)

1. ALL DRAFT ANGLES TO BE 2°.
2. ALL UNSPECIFIED RADII TO BE .015 R.
3. TYPICAL WALL THICKNESS .075".
4. EJECTOR PIN MARKS TO BE FLUSH TO .015 BELOW SURFACE.
5. DRAFT ANGLES ARE TO BE WITHIN DIMENSIONAL LIMITS ON THESE SURFACES. (NO DRAFT)
6. DIMENSIONS ARE TO BE – DRAFT.
7. EXTERIOR SURFACES TO HAVE A LIGHT MATTE FINISH: #RE 6622 (.00075 DP. @ 1.5° MINIMUM DRAFT)

COMPUTER-AIDED DRAFTING AND DESIGN	DRAWN BY	DATE	DWG NO.

DRAWN BY
DATE
DWG NO.

Make a working drawing of the *Welded Pulley Housing.* Use starting point A. Scale 1/4″= 1″. Convert fractions to decimals. Use current ANSI standards throughout.

BASIC TYPES OF JOINTS

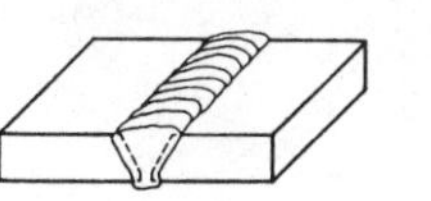
BUTT JOINT (V-GROOVE WELD)

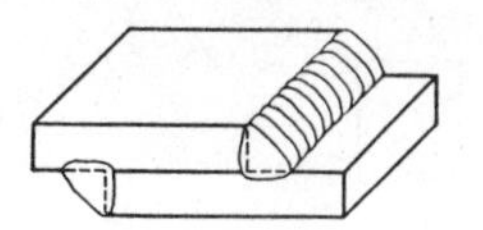
LAP JOINT (FILLET WELD)

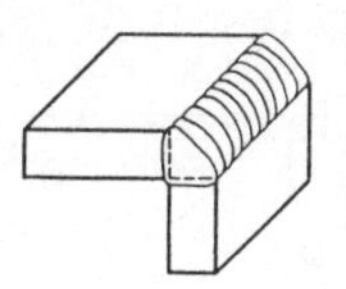
CORNER JOINT (FILLET WELD)

(FOR REFERENCE ONLY)

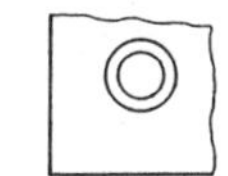
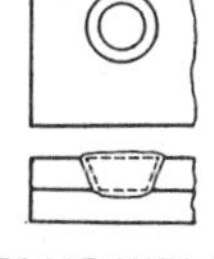
PLUG WELD

EDGE JOINT (V-GROOVE WELD)

T-JOINT (FILLET WELD)

SLOT WELD

A+

WELDING DRAFTING	DRAWN BY	DATE	DWG NO.

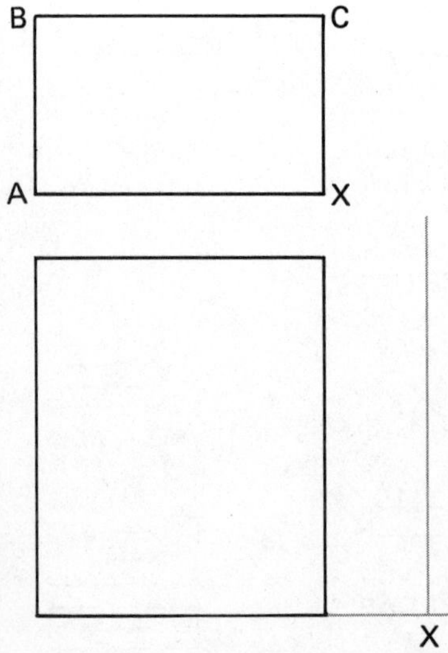

B C

A X

X

Using the front and top views shown at the left, develop a stretchout (pattern) for the rectangular prism. Add the top in its correct position for fabrication. Also add .08" tabs for fastening corners and mark bend lines. Do not dimension.

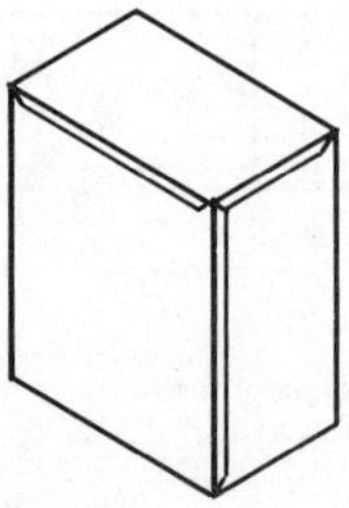

A

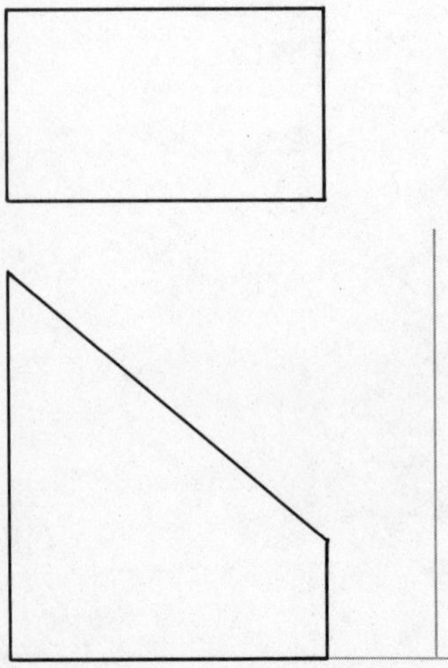

Using the front and top views shown at the left, develop a stretchout (pattern) for the truncated prism. Add the top in its correct position for fabrication. Also add .08" tabs for fastening corners and mark bend lines. Do not dimension.

B

Using the front and top views shown at the left, develop a stretchout (pattern) for the letter "N." Add the top (shape of the "N" in its correct position for fabrication. Also add .08" tabs for fastening corners and mark bend lines. Do not dimension.

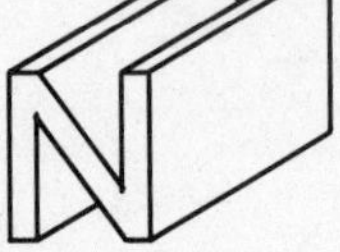

C

SURFACE DEVELOPMENT	DRAWN BY	DATE	DWG NO.

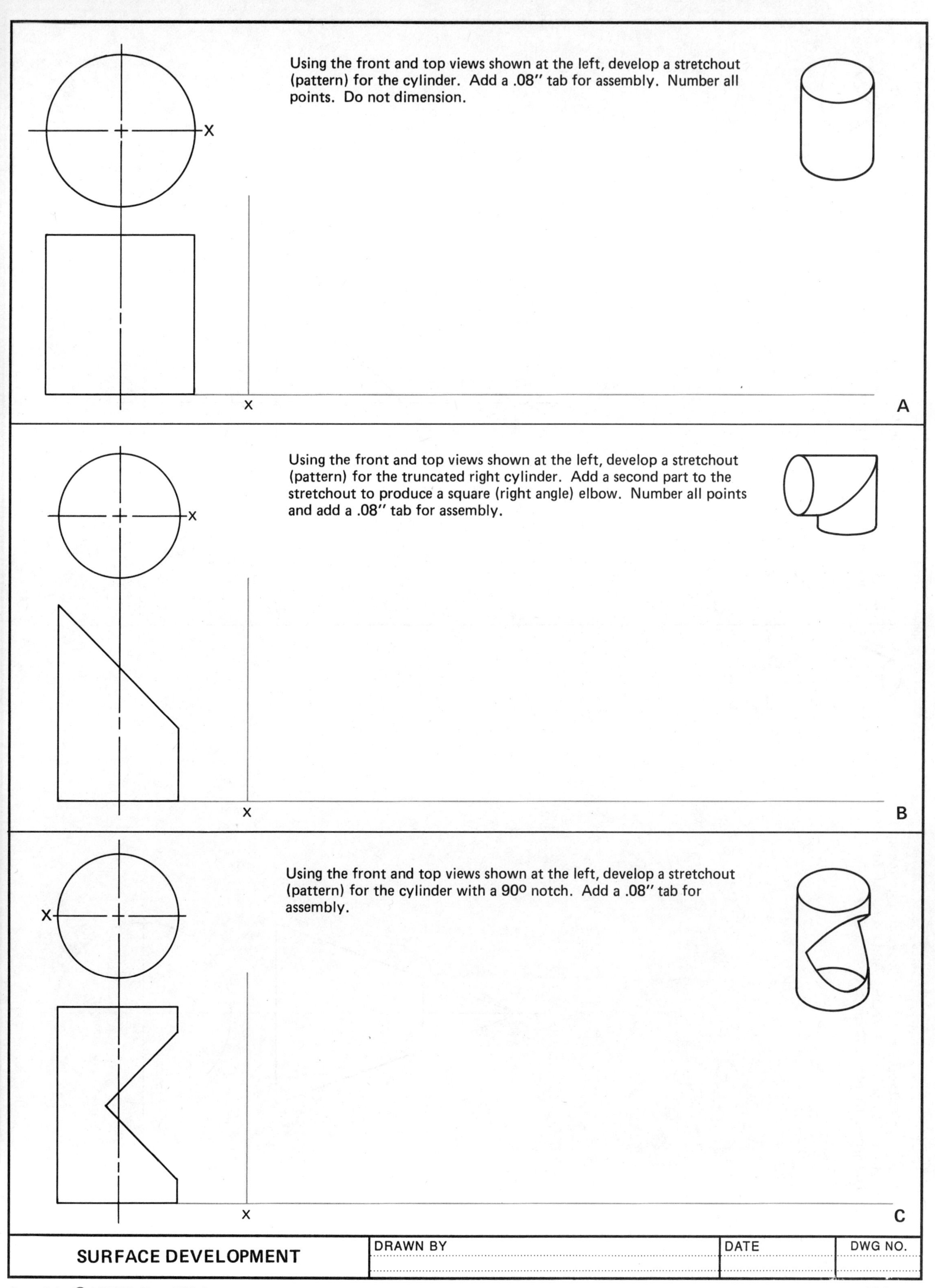

Using the front and top views shown at the left, develop a stretchout (pattern) for the cylinder. Add a .08" tab for assembly. Number all points. Do not dimension.

Using the front and top views shown at the left, develop a stretchout (pattern) for the truncated right cylinder. Add a second part to the stretchout to produce a square (right angle) elbow. Number all points and add a .08" tab for assembly.

Using the front and top views shown at the left, develop a stretchout (pattern) for the cylinder with a 90° notch. Add a .08" tab for assembly.

SURFACE DEVELOPMENT	DRAWN BY	DATE	DWG NO.

Using the front and top views shown, develop a stretchout (pattern) for the frustum of a right circular cone. Add a .08" tab for assembly. Do not dimension. Number all points.

A

Using the top and front views shown, develop a stretchout (pattern) for the truncated right circular cone. Add a .08" tab for assembly. Do not dimension. Number all points.

B

SURFACE DEVELOPMENT

DRAWN BY	DATE	DWG NO.

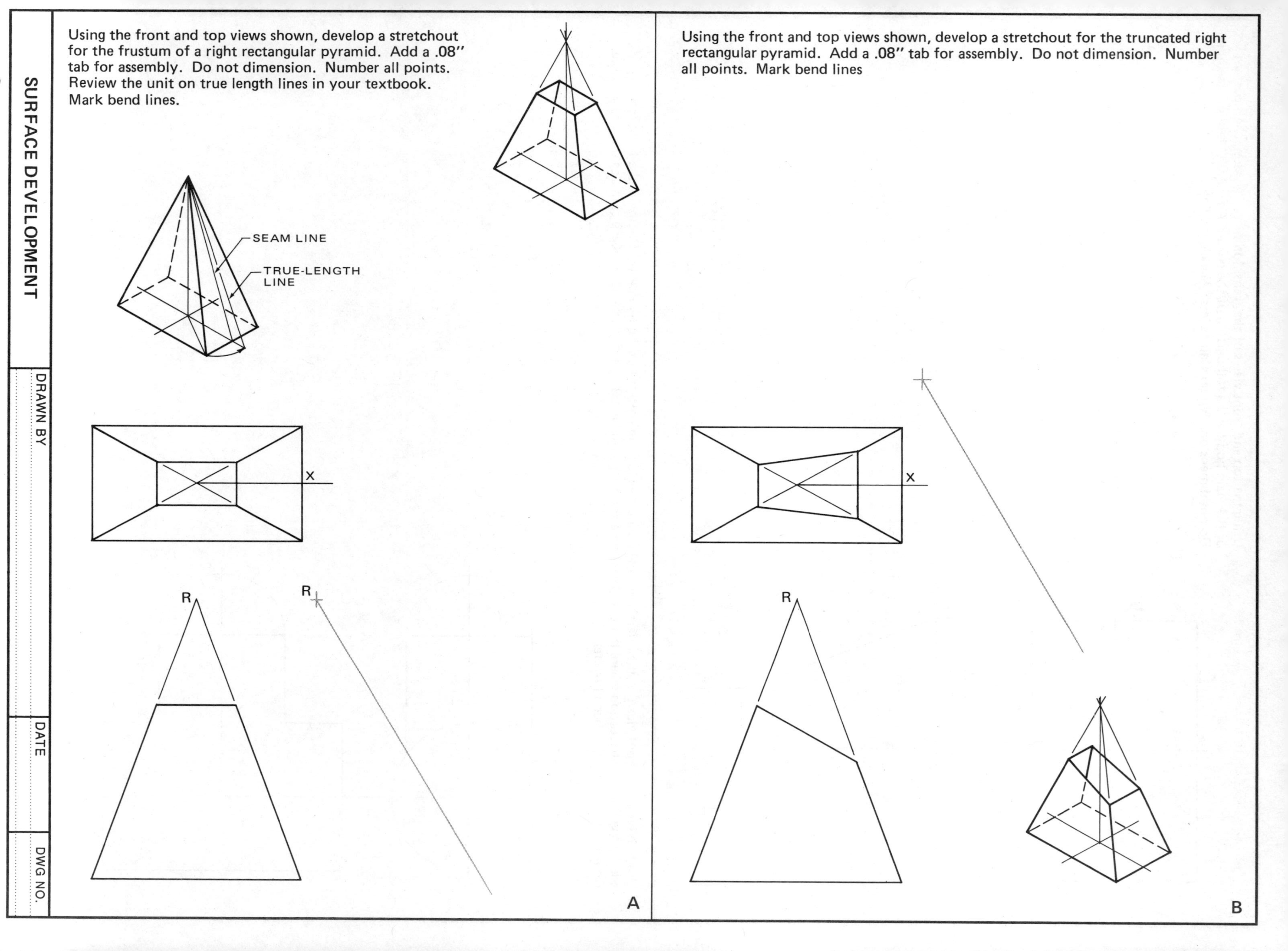
Using the front and top views shown, develop a stretchout for the frustum of a right rectangular pyramid. Add a .08" tab for assembly. Do not dimension. Number all points. Review the unit on true length lines in your textbook. Mark bend lines.
SEAM LINE
TRUE-LENGTH LINE
X
R
R
A
Using the front and top views shown, develop a stretchout for the truncated right rectangular pyramid. Add a .08" tab for assembly. Do not dimension. Number all points. Mark bend lines
X
R
B
SURFACE DEVELOPMENT
DRAWN BY
DATE
DWG NO.
Copyright © 1990 McGraw-Hill, Inc.
18-4

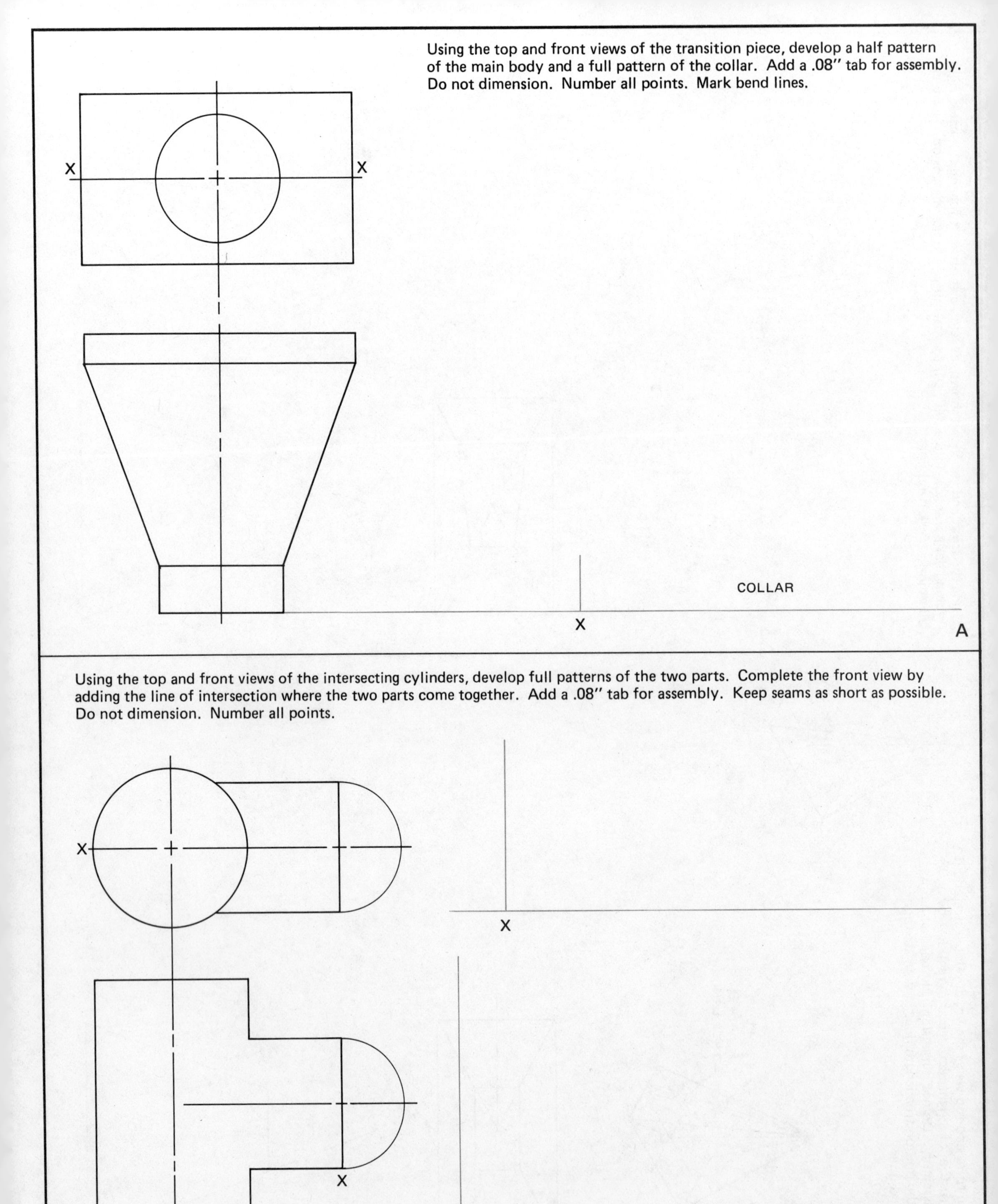

SURFACE DEVELOPMENT	DRAWN BY	DATE	DWG NO.

SPUR GEARS

Study the unit on gears in your textbook. Use the information given below and calculate all other data necessary to draw the spur gear shown at the right. Estimate all sizes and details not given. Scale: Half size.

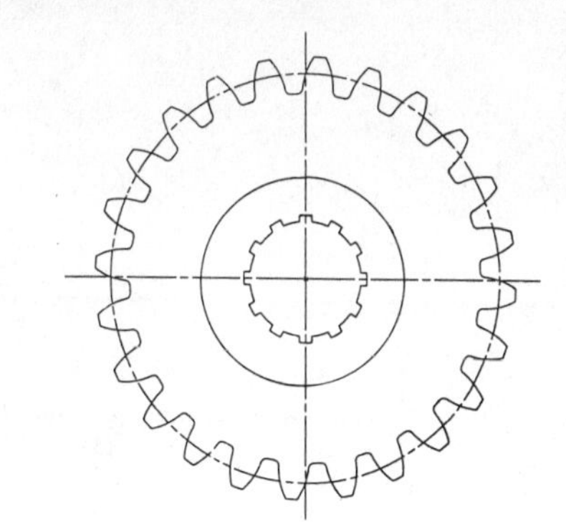

TOOTH DATA

DIAMETRAL PITCH – 3
PRESSURE ANGLE – 20°
NUMBER OF TEETH – 24
PITCH DIAMETER – 8
GEAR THICKNESS – 1.25

HUB DATA

HUB DIAMETER – 2.00
HOLE DIAMETER – 1.125
SPLINE – .25 X .125

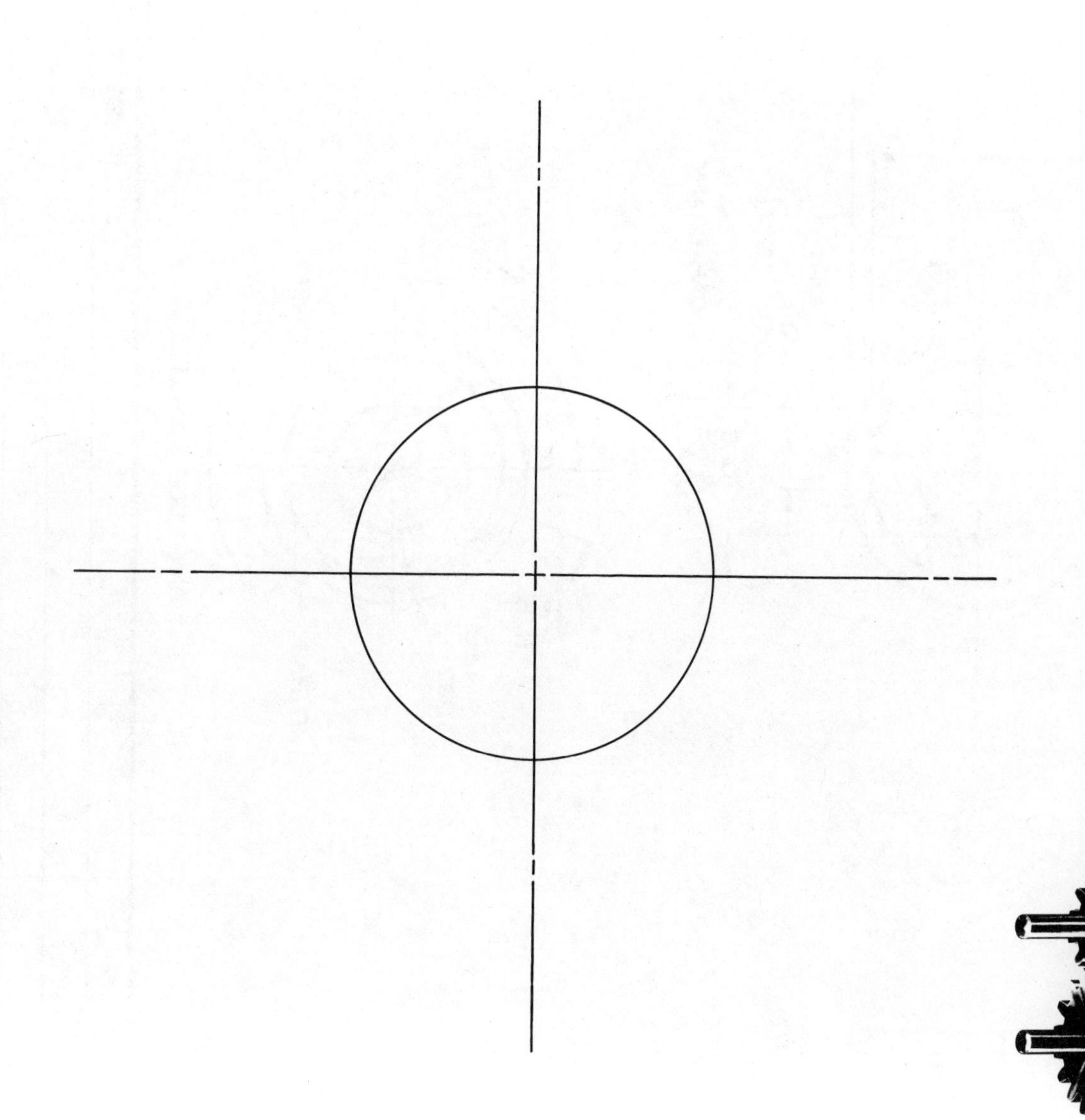

GEARS	DRAWN BY	DATE	DWG NO.

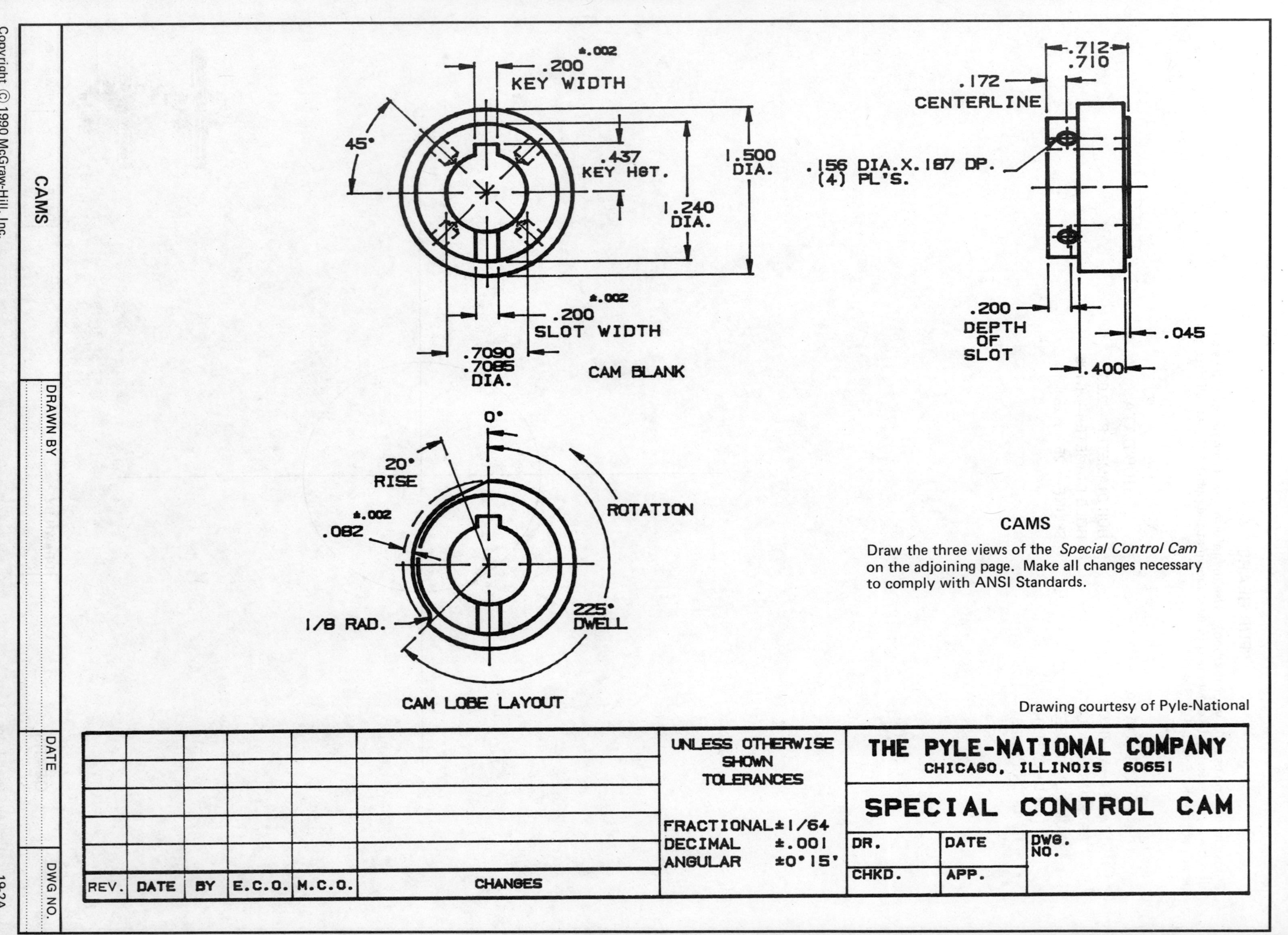

CAMS

Draw the three views of the *Special Control Cam* on the adjoining page. Make all changes necessary to comply with ANSI Standards.

Drawing courtesy of Pyle-National

CAMS

DRAWN BY

DATE

DWG NO.

CAMS	DRAWN BY	DATE	DWG NO.

NOTATION AND DETAIL

ROOFING
ROOF SLOPE
ROOF CONSTRUCTION
FLASHING
PARAPET OR OVERHANG DIMENSIONS
SOFFIT
SOFFIT/ATTIC VENTILATION
CEILING CONSTRUCTION
EXTERIOR WALL CONSTRUCTION
EXTERIOR WALL FLASHING/INSULATION
INTERIOR WALL CONSTRUCTION/FINISH
HEADER TYPE
SILL TYPE
HEADER/SILL HEIGHTS
SECOND FLOOR CONSTRUCTION
GROUND FLOOR EXTERIOR WALL
FIRST FLOOR CONSTRUCTION
FOOTING/FOUNDATION WALL
BASEMENT WALL
BASEMENT FLOOR/FOOTING
WATERPROOFING

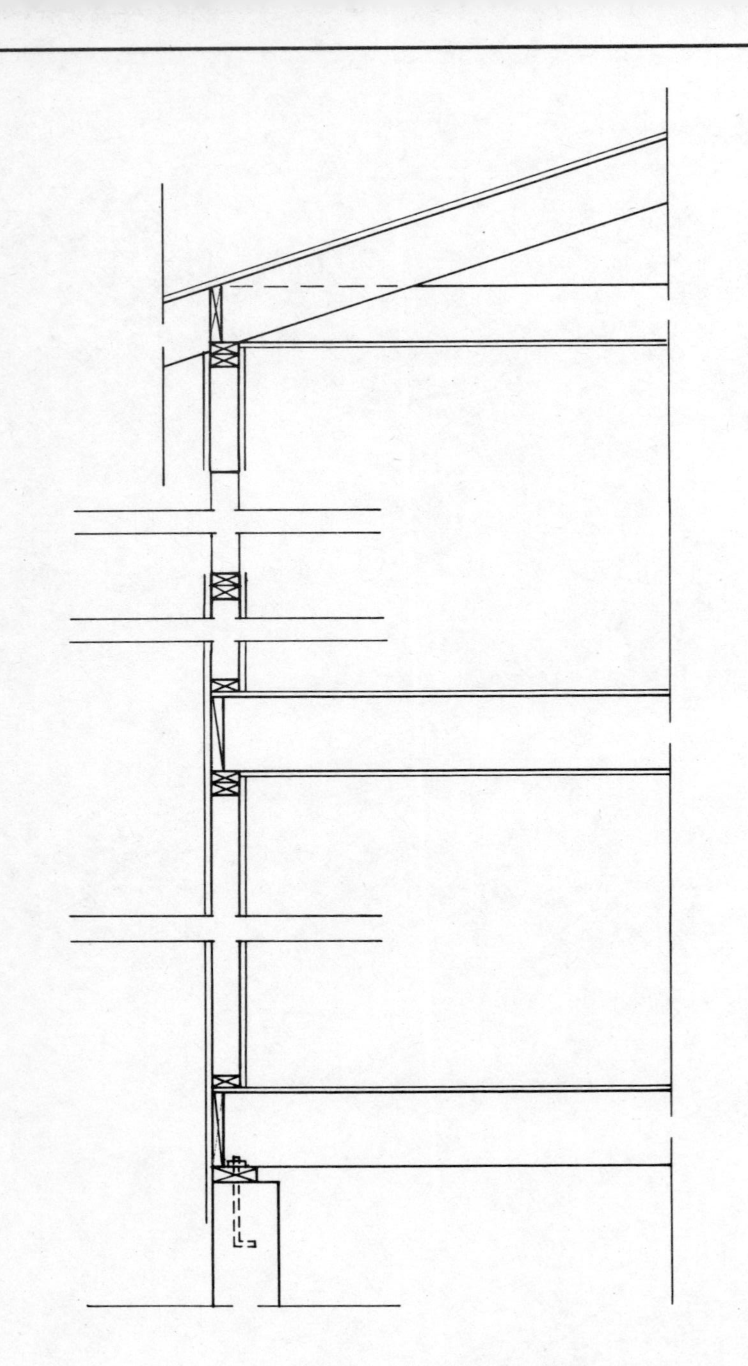

Complete the platform wall section. Label all details with an architectural-style lettering. Include title and scale (1/2″ = 1′–0″). Some typical notations are provided above.

ARCHITECTURAL DRAFTING	DRAWN BY	DATE	DWG NO.

NOTATION AND DETAIL

ROOFING
ROOF SLOPE
ROOF CONSTRUCTION
FLASHING
PARAPET OR OVERHANG DIMENSIONS
SOFFIT
SOFFIT/ATTIC VENTILATION
CEILING CONSTRUCTION
EXTERIOR WALL CONSTRUCTION
EXTERIOR WALL FLASHING/INSULATION
INTERIOR WALL CONSTRUCTION/FINISH
HEADER TYPE
SILL TYPE
HEADER/SILL HEIGHTS
SECOND FLOOR CONSTRUCTION
GROUND FLOOR EXTERIOR WALL
FIRST FLOOR CONSTRUCTION
FOOTING/FOUNDATION WALL
BASEMENT WALL
BASEMENT FLOOR/FOOTING
WATERPROOFING

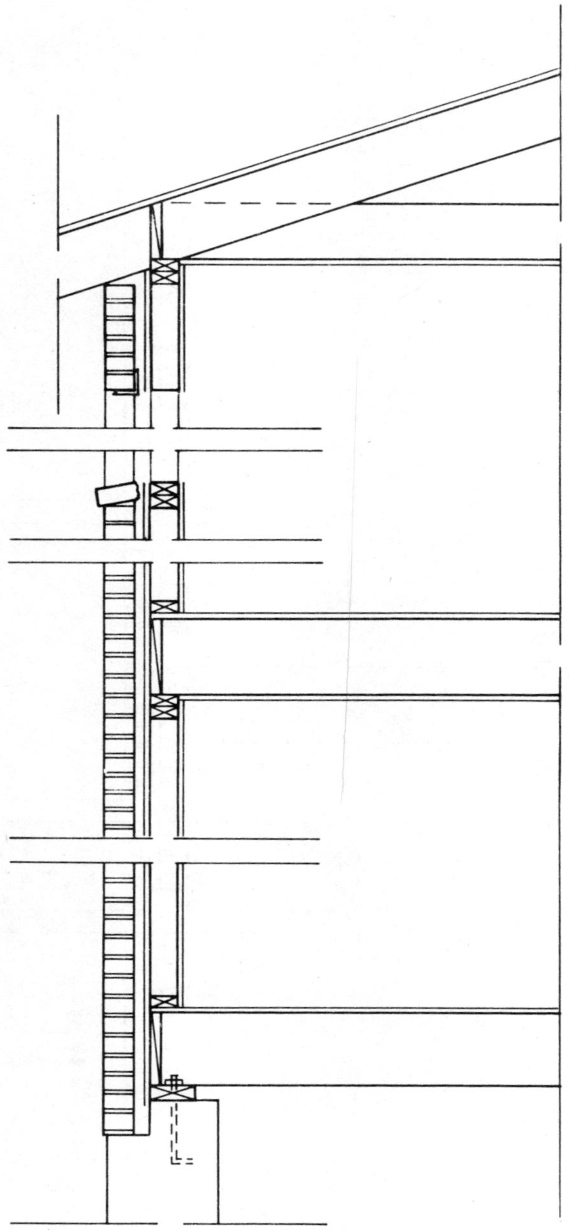

Complete the brick veneer wall section. Label all details with an architectural-style lettering. Include title and scale (1/2" = 1'–0"). Some typical notations are provided above.

ARCHITECTURAL DRAFTING	DRAWN BY	DATE	DWG NO.

NOTATION AND DETAIL

ROOFING
ROOF SLOPE
ROOF CONSTRUCTION
FLASHING
PARAPET OR OVERHANG DIMENSIONS
SOFFIT
SOFFIT/ATTIC VENTILATION
CEILING CONSTRUCTION
EXTERIOR WALL CONSTRUCTION
EXTERIOR WALL FLASHING/INSULATION
INTERIOR WALL CONSTRUCTION/FINISH
HEADER TYPE
SILL TYPE
HEADER/SILL HEIGHTS
SECOND FLOOR CONSTRUCTION
GROUND FLOOR EXTERIOR WALL
FIRST FLOOR CONSTRUCTION
FOOTING/FOUNDATION WALL
BASEMENT WALL
BASEMENT FLOOR/FOOTING
WATERPROOFING

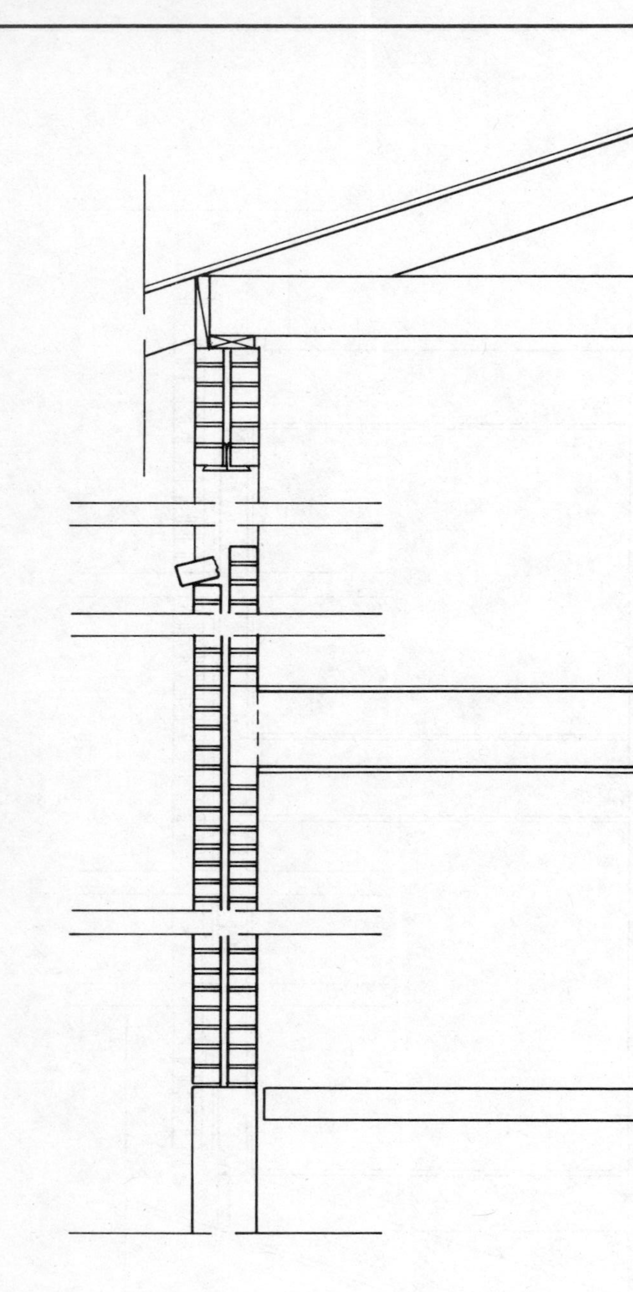

Complete the 8″ brick wall section. Label all details with an architectural-style lettering. Include title and scale (1/2″ = 1′–0″). Some typical notations are provided above.

ARCHITECTURAL DRAFTING	DRAWN BY	DATE	DWG NO.

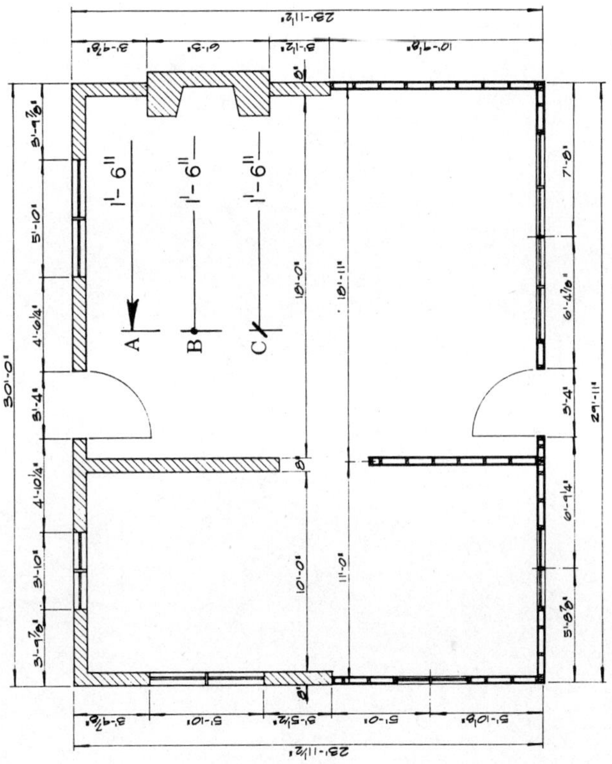

Complete the floor plan shown. Add title, scale, and all dimensions. Give careful attention to architectural drafting style and techniques.

ARCHITECTURAL DRAFTING	DRAWN BY	DATE	DWG NO.

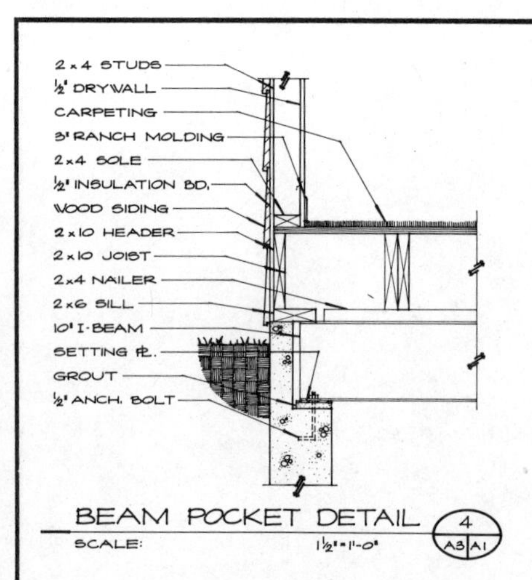

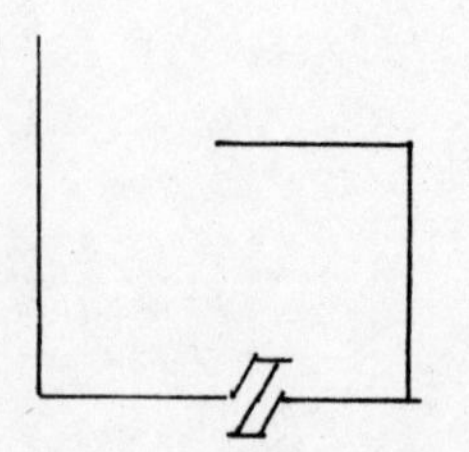

Draw the wall section shown above. Add a footing designed for your geographic location. Scale: 1-1/2" = 1'– 0".

ARCHITECTURAL DRAFTING	DRAWN BY	DATE	DWG NO.

Duplicate the drawing and information below for practice in architectural lettering and design detailing.

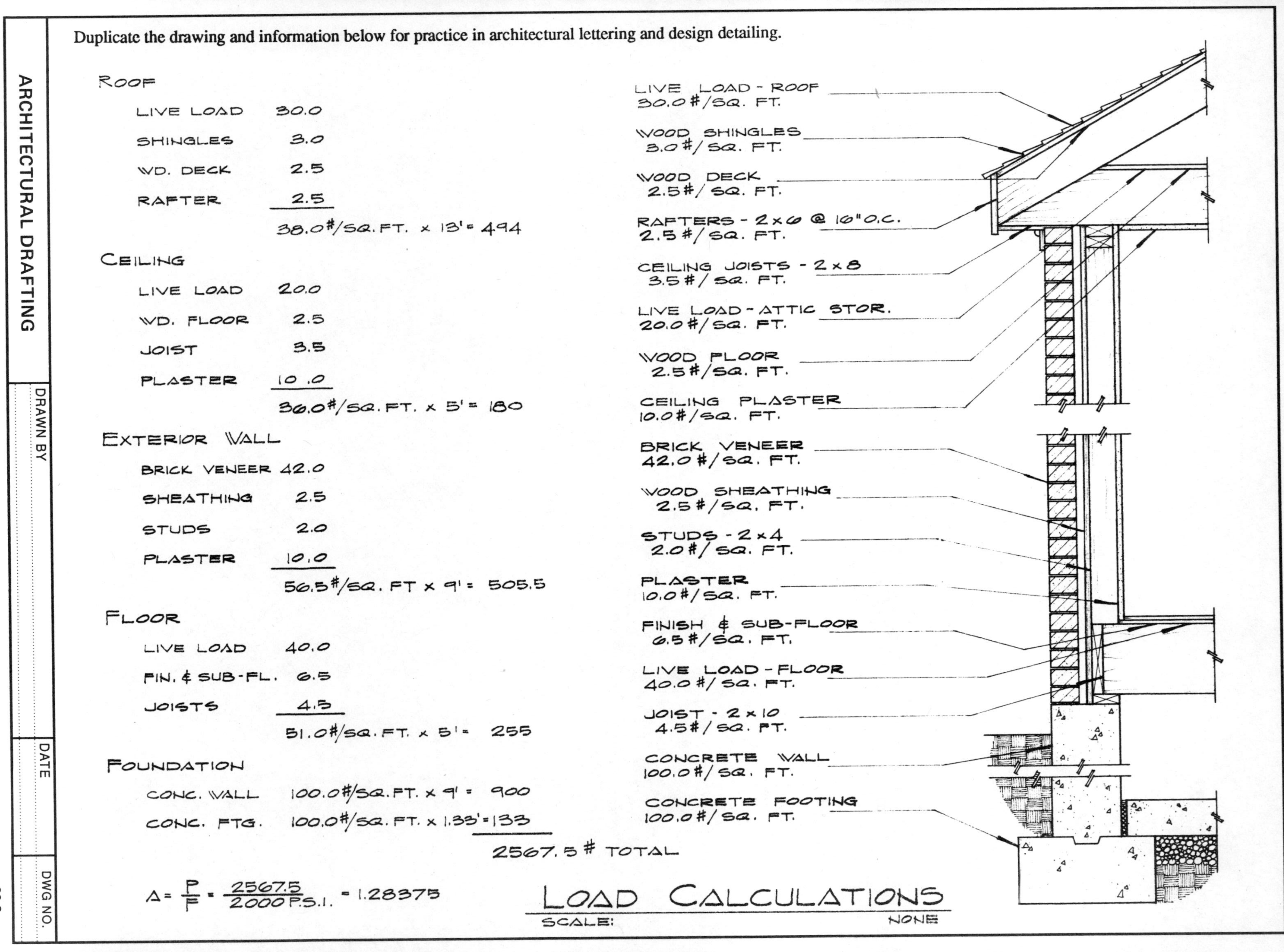

	DRAWN BY	DATE	DWG NO.

On separate sheets, create a preliminary study (example below) of a bachelor's retreat. Use a scale of 1/16″ = 1′– 0″ for the floor plan and plot plan and 1/8″ = 1′– 0″ for the elevation. Study the example carefully before you begin this assignment.

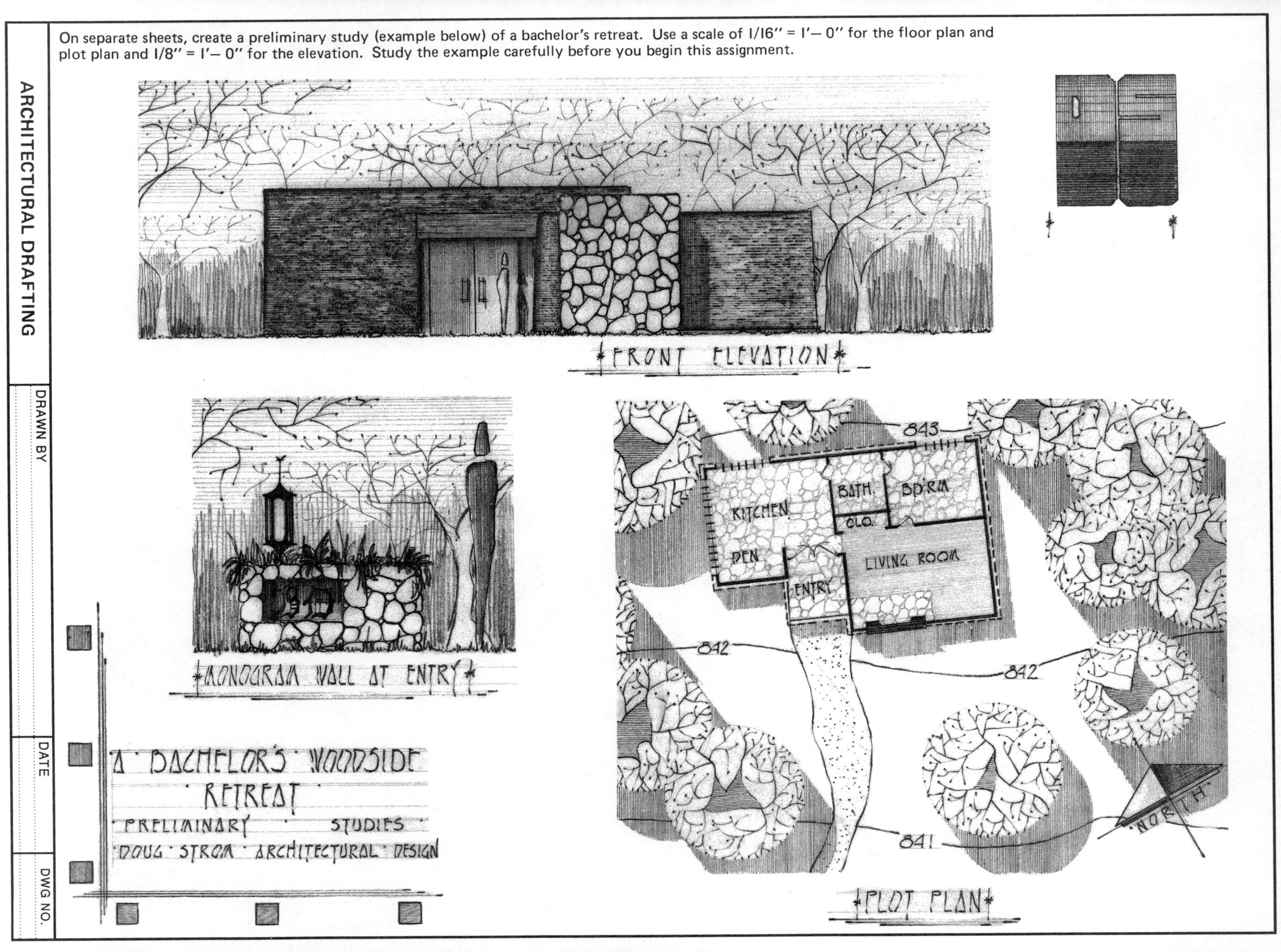

ARCHITECTURAL DRAFTING

DRAWN BY	DATE	DWG NO.

DRAWN BY
DATE
DWG NO.

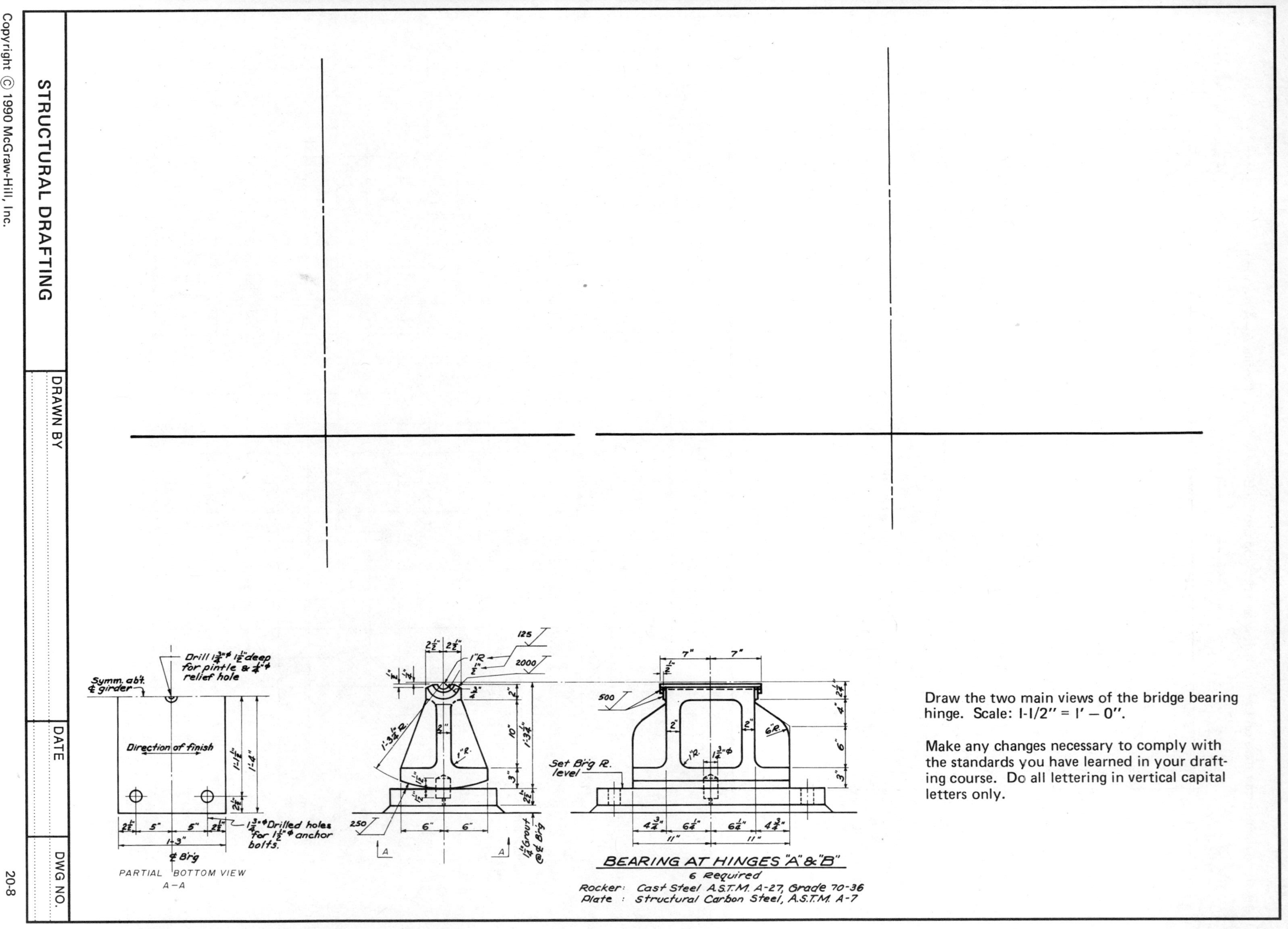

Draw the two main views of the bridge bearing hinge. Scale: 1-1/2″ = 1′ – 0″.

Make any changes necessary to comply with the standards you have learned in your drafting course. Do all lettering in vertical capital letters only.

STRUCTURAL DRAFTING	DRAWN BY	DATE	DWG NO.

Redraw the contour map on the grid below and develop profile B–B or C–C. Profile A–A is shown.

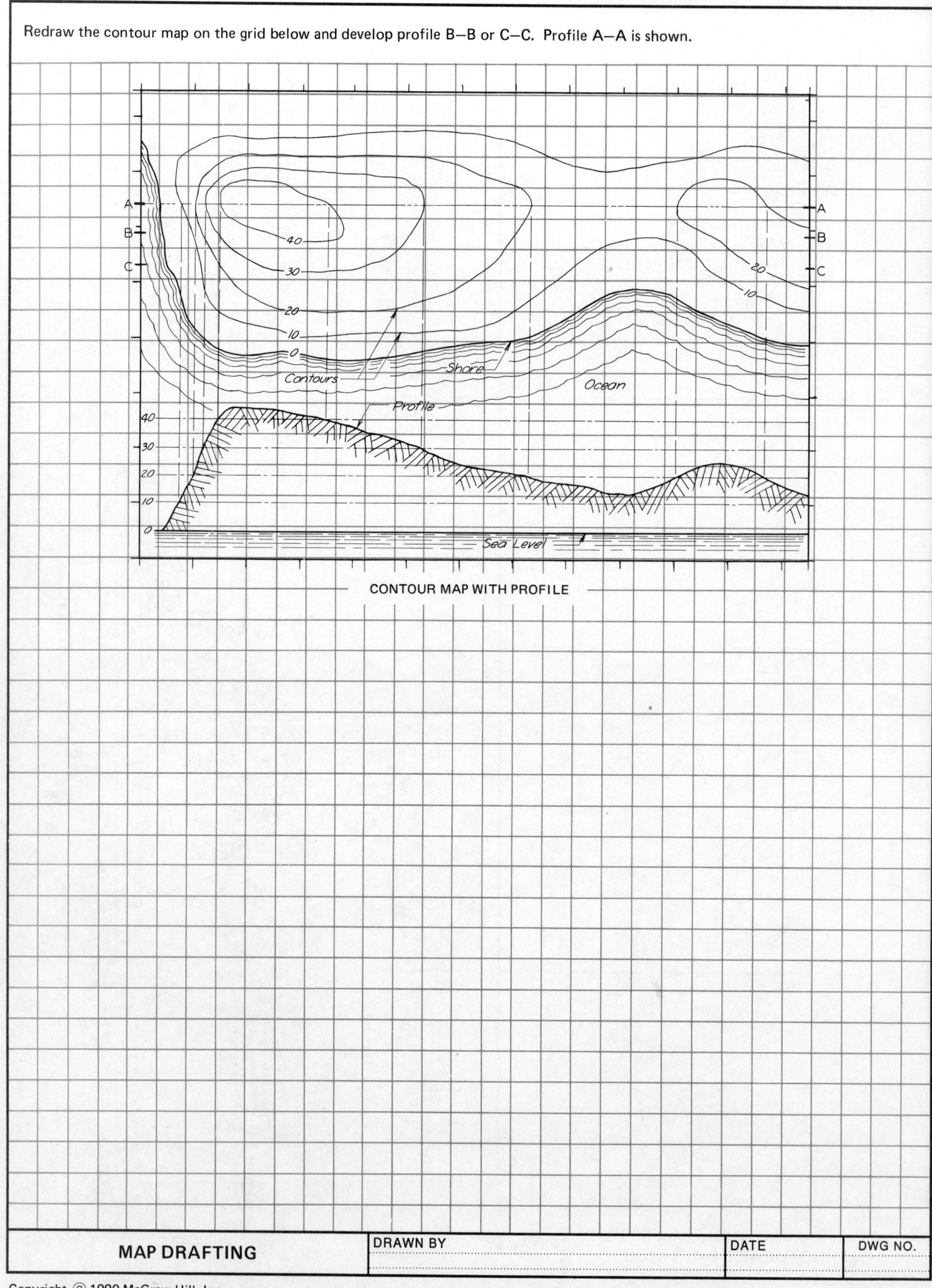

CONTOUR MAP WITH PROFILE

MAP DRAFTING	DRAWN BY	DATE	DWG NO.

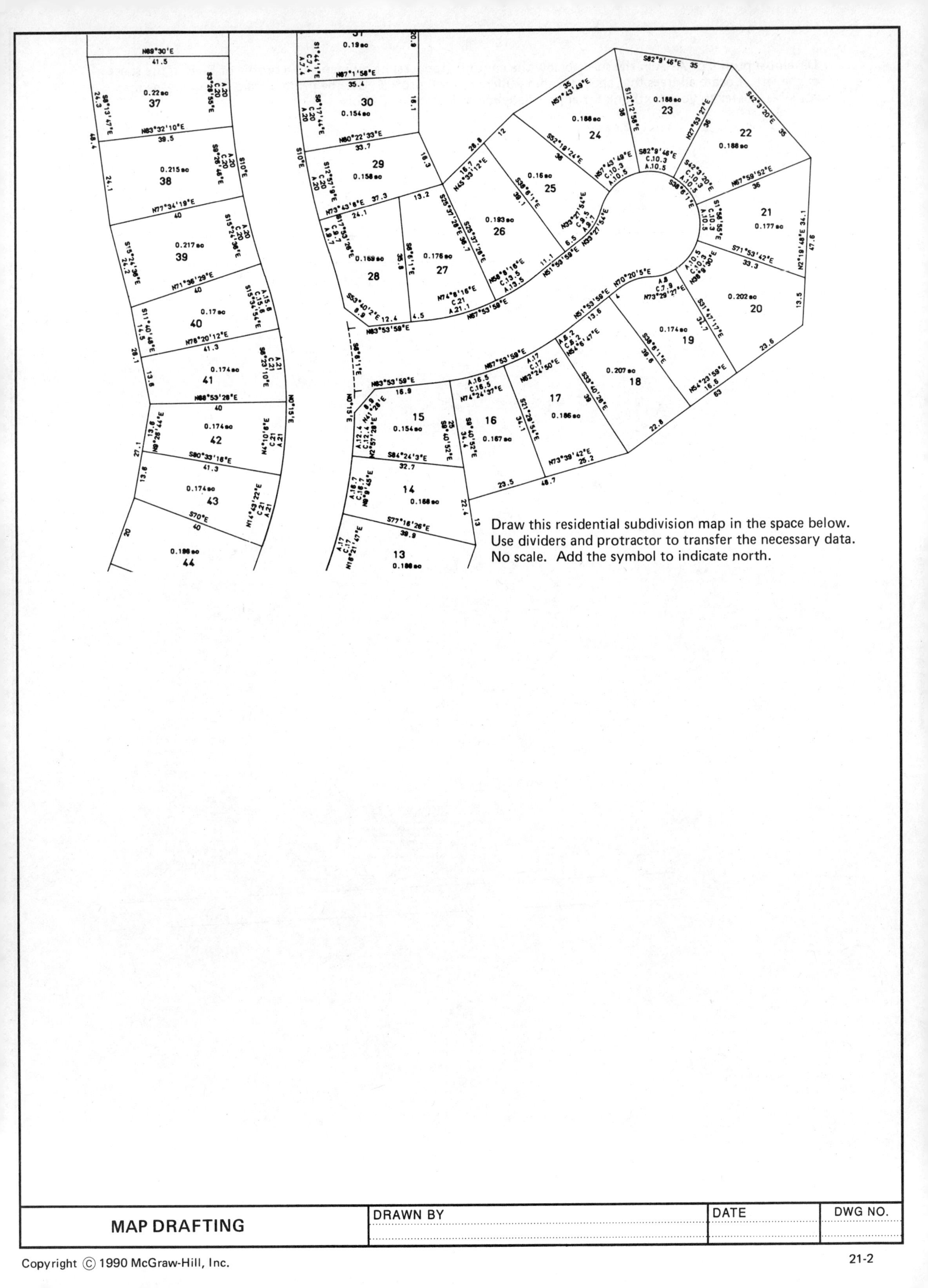

Draw this residential subdivision map in the space below. Use dividers and protractor to transfer the necessary data. No scale. Add the symbol to indicate north.

MAP DRAFTING	DRAWN BY	DATE	DWG NO.

Develop a profile of A–A in the space below the contour (topographical) map and a profile of B–B in the space at the left. Locate all trees that appear within 20 feet of each profile. Use the printed scale shown on the map for sizes. Refer to your textbook for appropriate line weights and symbols.

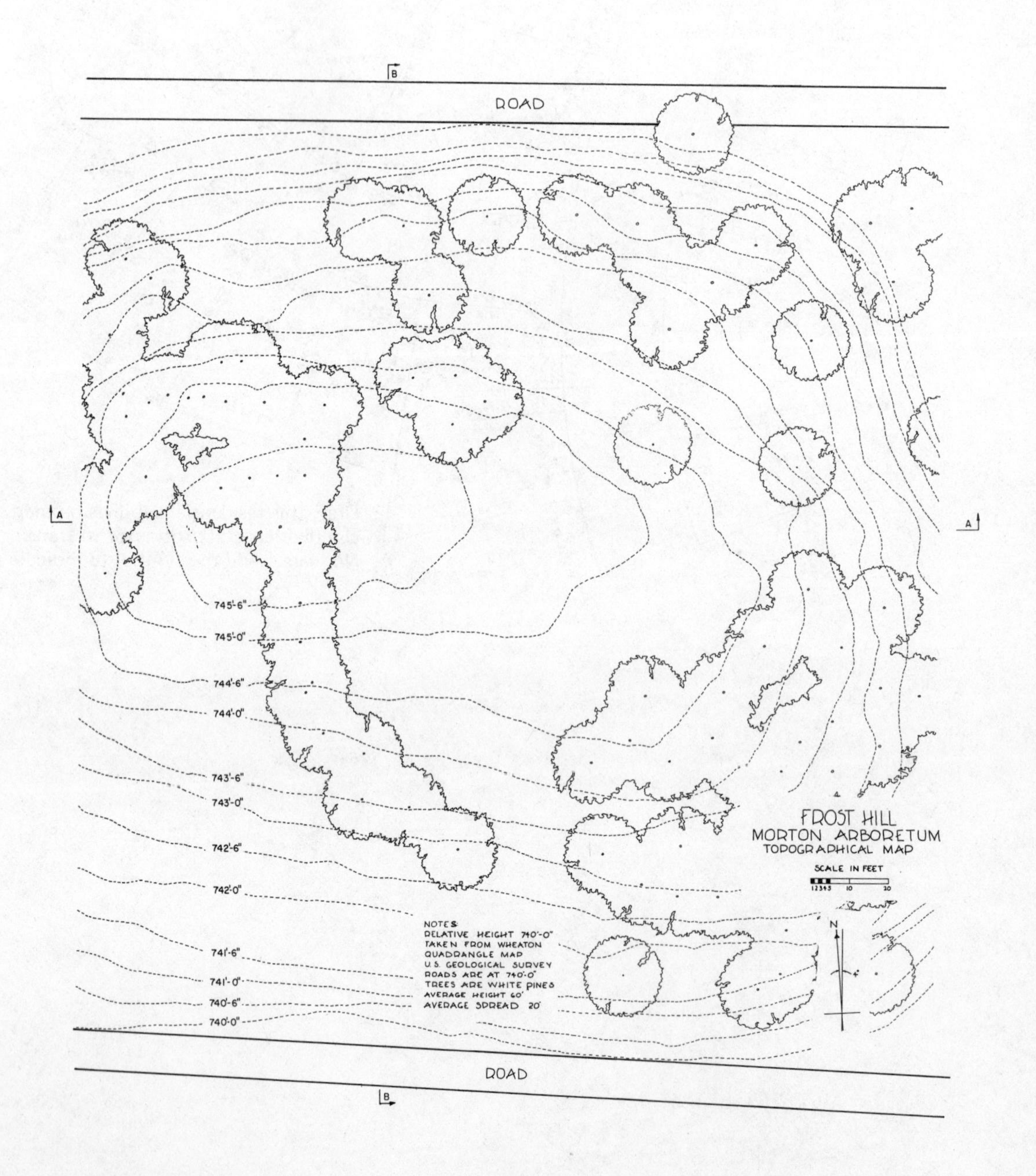

MAP DRAFTING	DRAWN BY	DATE	DWG NO.

Use the circle at the right to develop a pie chart showing how you use your time during an average weekday. Each hour equals 15° on the circle. Estimate fractional parts of an hour. Letter your name in the space below the chart title.

Begin by listing all activities and estimate the time you spend in each. When you are satisfied with your list, transfer the information to the pie chart.

Letter the activity, time, and percentage in each segment.

ACTIVITY	TIME	%

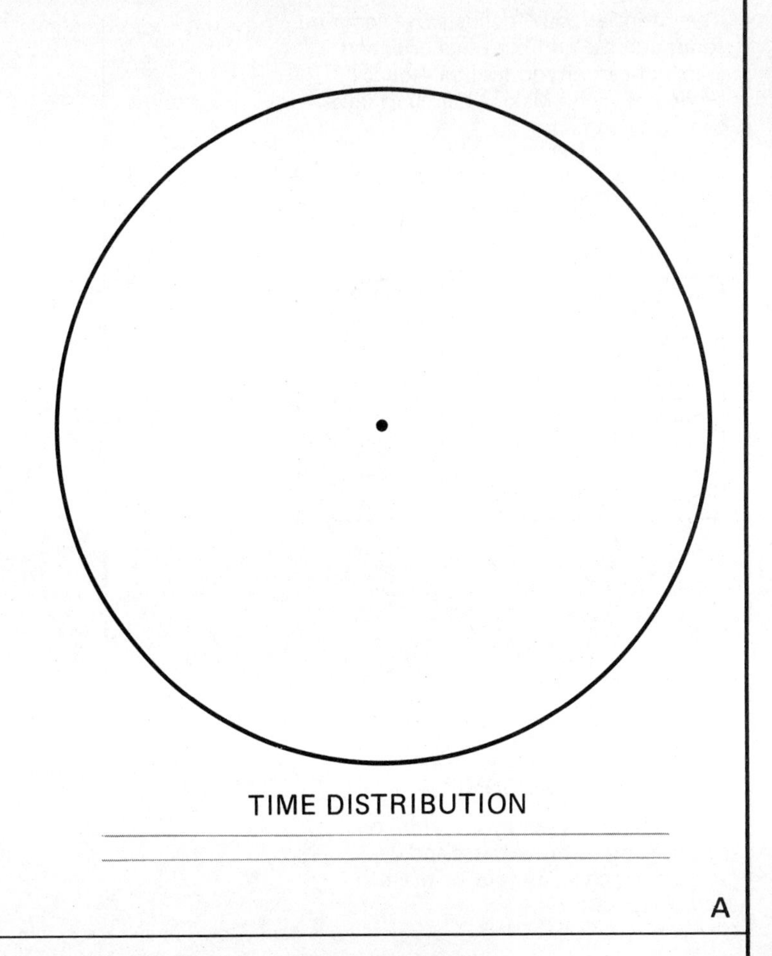

TIME DISTRIBUTION

A

Make a multiline chart to show scoring information on a school basketball team using the data in the table below. Use different colors or different line styles for each year. Include a key on or near the chart showing which line style or color is used for each year.

SCORING INFORMATION			
GAME	1988	1989	1990
1	46	55	62
2	64	60	55
3	52	55	60
4	35	50	58
5	47	52	70
6	39	48	62
7	42	55	60
8	57	50	65
9	72	50	45
10	82	54	75

POINTS

BASKETBALL GAMES

ALTERNATE ASSIGNMENT:

Obtain information on rainfall, temperature, etc. Develop multiline charts to show comparisons on a week-to-week, month-to-month, or year-to-year basis.

B

GRAPHIC CHARTS AND DIAGRAMS	DRAWN BY	DATE	DWG NO.

The information below lists five common foods and the number of calories and grams of carbohydrates in a 4-ounce serving of each. Make a bar chart illustrating these facts.

FOOD	CALORIES	CARBO-HYDRATES
Chocolate ice cream	150	14
Peas	75	14
Pizza	260	29
Milk	85	6
Strawberries	30	6

ALTERNATE ASSIGNMENT:

Find a list of similar information on calories and carbohydrates and develop a bar chart on four or five of *your* favorite foods.

A

Draw a vertical-bar chart showing the following student attendance for a given week of school. The total school enrollment is 925.

DAYS	ATTENDANCE
Monday	625
Tuesday	715
Wednesday	800
Thursday	775
Friday	695

ALTERNATE ASSIGNMENT:

Obtain similar data for your school, church, club, etc. Keep a weekly or monthly set of charts on attendance. Display them to show trends in attendance and enrollment.

B

GRAPHIC CHARTS AND DIAGRAMS	DRAWN BY	DATE	DWG NO.

Examine the schematic diagram for an *AC SOUND-OPERATED SWITCH.* The switch operates with the clap of your hand. Draw the diagram below twice the size shown. Use dividers or scale and electronic symbol template.

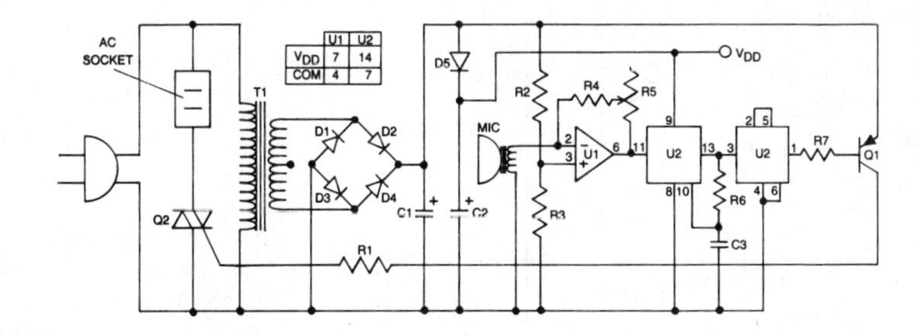

	U1	U2
V_{DD}	7	14
COM	4	7

ELECTRICAL AND ELECTRONICS DRAFTING	DRAWN BY	DATE	DWG NO.

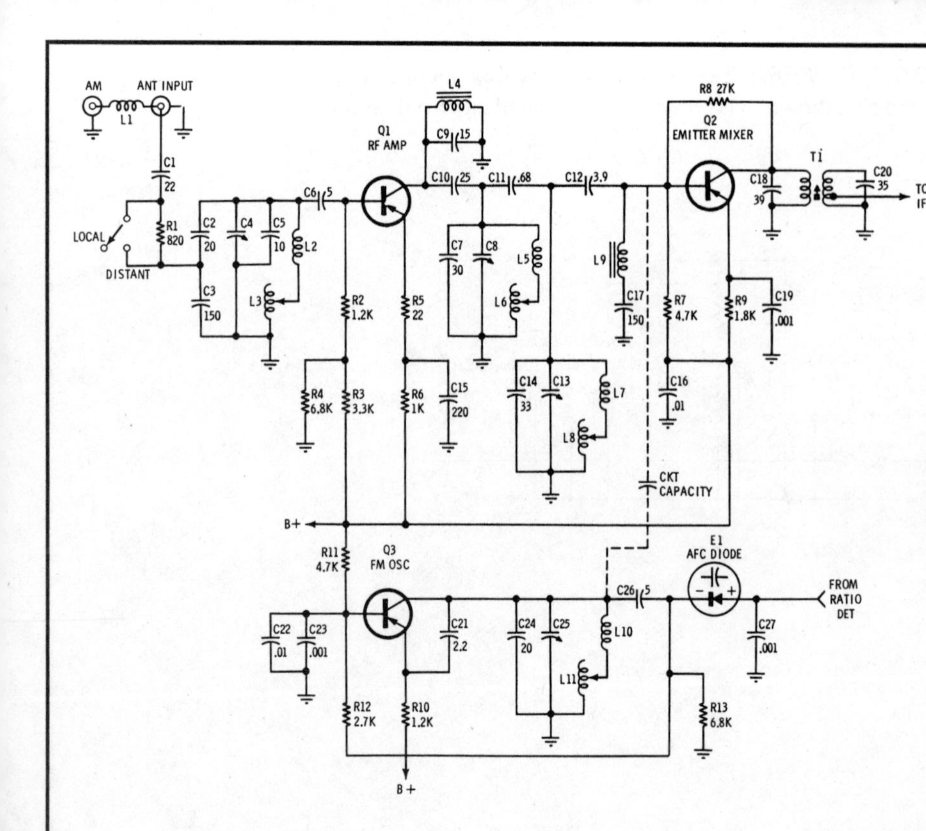

At the bottom of this sheet or on a separate sheet, draw the schematic diagram for a tuner used in an FM stereo auto radio. Prepare a parts list alphabetically. Use a template for electronic symbols.

ELECTRICAL AND ELECTRONICS DRAFTING	DRAWN BY	DATE	DWG NO.

Make a working drawing of the *Landing Gear and Lift Strut Fitting* with the following changes. The V-tongue is 1-1/4" instead of 1-9/16". The 1-3/16" location dimension is to be 1-3/4". Note that the 1/8" label applies to the V-tongue and other radii. SCALE: 3/4"=1".

OPTIONAL ASSIGNMENTS: List the sequence of operations necessary to make this fitting. Change all fractions to decimals. Complete this assignment using your CAD system.

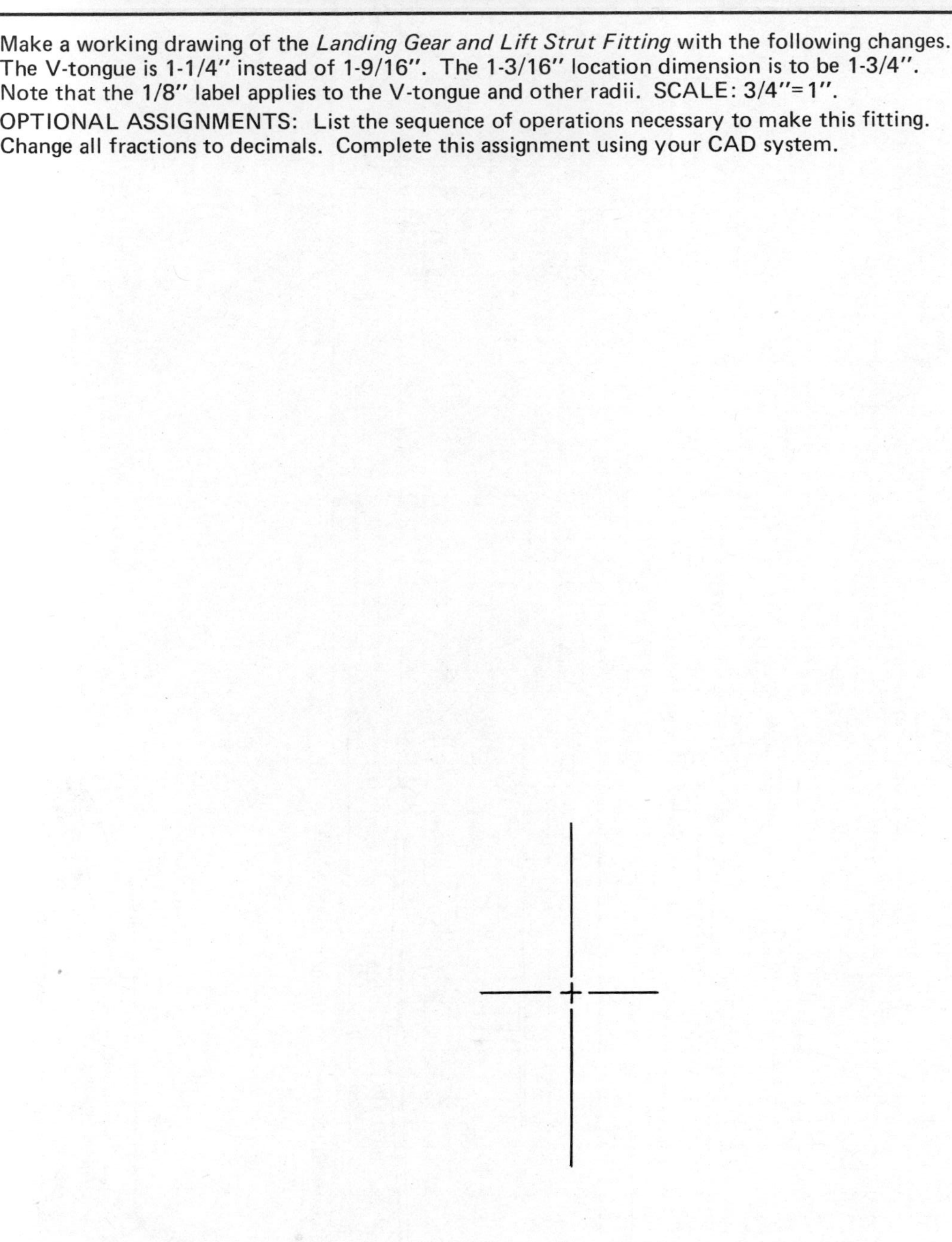

AEROSPACE DRAFTING

DRAWN BY

DATE

DWG NO.

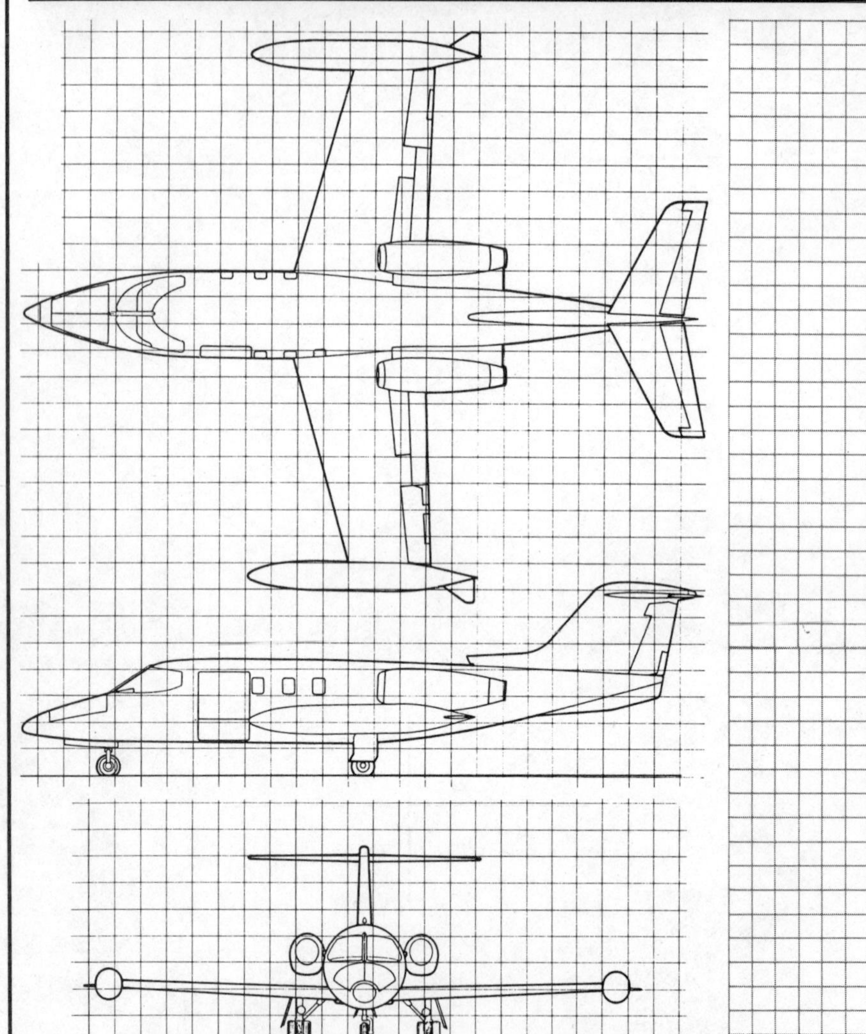

Sketch the three views on the grid at the right.

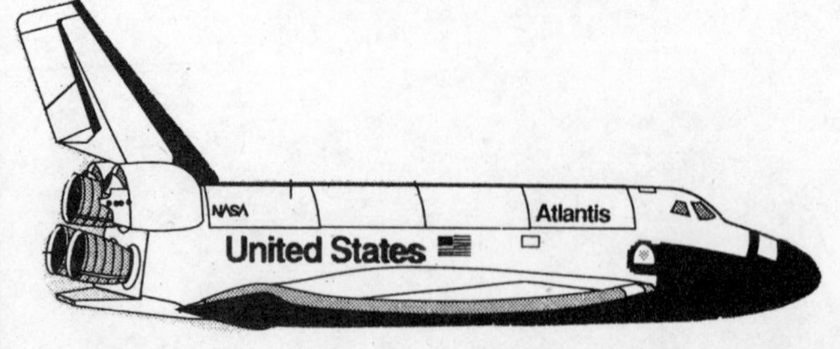

Sketch the profile of the *Space Shuttle* in the open space at the right.

AEROSPACE DRAFTING	DRAWN BY	DATE	DWG NO.

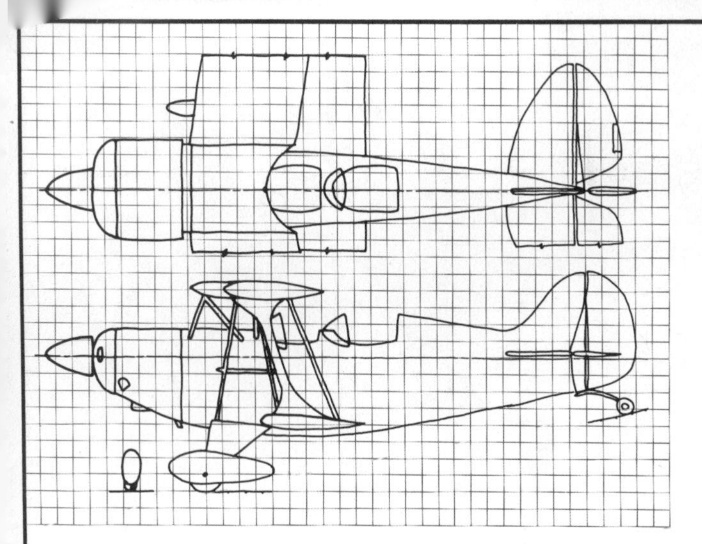

Sketch the *Pitts Aviation Aerobotic Biplane* on the grid below. Feel free to make small changes in the design.

OPTIONAL ASSIGNMENT: After completing the sketch, place a sheet of tracing media over it and use irregular curves to refine the profile.

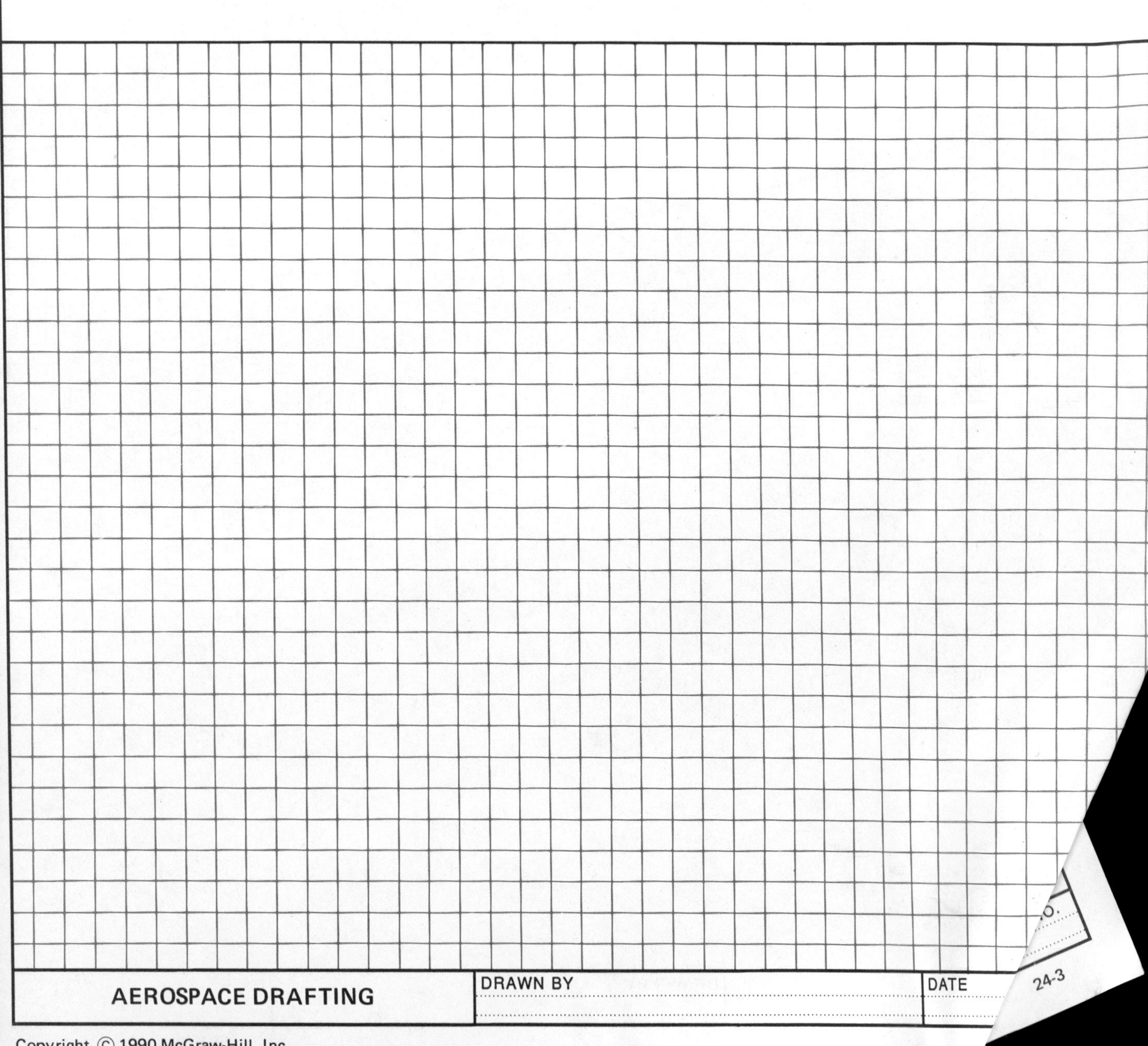

AEROSPACE DRAFTING	DRAWN BY	DATE	24-3

DRAWN BY
DATE
DWG NO.
Copyright © 1990
Hill, Inc.